An Abundance of Curiosities

An Abundance

of Curiosities

The Natural History of North Carolina's Coastal Plain

ERIC G. BOLEN AND JAMES F. PARNELL

The University of Georgia Press ATHENS

SOUTHERN
HIGHLANDS
RESERVE
Publications

© 2022 by the University of Georgia Press
Athens, Georgia 30602
www.ugapress.org
All rights reserved
Designed by Erin Kirk
Set in Miller Text
Printed and bound by Versa Press

The paper in this book meets the guidelines for
permanence and durability of the Committee on
Production Guidelines for Book Longevity of the
Council on Library Resources.

Most University of Georgia Press titles are
available from popular e-book vendors.

Printed in the United States of America
26 25 24 23 22 C 5 4 3 2 1

Library of Congress Cataloging-in-Publication Data

Names: Bolen, Eric G., author. | Parnell, James F., author.
Title: An abundance of curiosities : the natural history of North Carolina's
    coastal plain / Eric G. Bolen and James F. Parnell.
Description: Athens : The University of Georgia Press, 2022. | Includes
    bibliographical references.
Identifiers: LCCN 2021058920 | ISBN 9780820361765 (hardback)
Subjects: LCSH: Natural history—North Carolina—Atlantic Coast. |
    Coasts—North Carolina. | Coastal ecology—North Carolina.
Classification: LCC QH105.N8 B65 2022 | DDC 508.756—dc23/eng/20211221
LC record available at https://lccn.loc.gov/202105892

To those who treasure and protect wild places everywhere.

Contents

Foreword

Even though I did not have the privilege of attending their classes, Eric Bolen and James Parnell—both with distinguished careers at the University of North Carolina Wilmington—have been teachers and mentors to me and numerous other naturalists and scientists. Their writing, scholarship, and leadership have helped shape the study of ecology in North Carolina and the nation. Bolen has written or coauthored important works on waterfowl, wildlife ecology and management, the ecology of North America, and, most recently, the natural history of Texas. James Parnell was a pioneer in the study of coastal birds in North Carolina and is one of the best-known writers and photographers on all things birds, mammals, and fish in the Carolinas. To see their talents come together in one volume, *An Abundance of Curiosities: The Natural History of North Carolina's Coastal Plain*, is a marriage made in ecological heaven.

Having been one of the first to savor this book, I am confident it will excite both accomplished scholars and casual students of the natural history of the Tar Heel coastal region. One of the most difficult tasks in writing is blending seemingly disparate fields—biology, coastal geology, evolution, botany, climate, and social history. This book is the evidence of their success. Bolen is a crisp writer whose expertise and passion for the land can be seen on every page. He is also intimately familiar with the chroniclers, explorers, and naturalists from North Carolina's rich coastal heritage. The photographs of James Parnell have long set the standard in bird and wildlife photography. Both are residents of coastal North Carolina, and their "sense of place" shines through in every chapter.

Although this book is designed for a general audience— vacationers, birders, students, wildflower lovers, teachers, and citizen scientists— please don't regard this as simply a "coastal guide." View it as a portal to a deeper understanding of the rich natural communities found between Interstate 95 and the Outer Banks. Bolen weaves the cultural history of the region and its key players—Native Americans, early European explorers, enslaved people, early naturalists, fisher people, and small farmers—with his deep knowledge of the ecology of coastal North Carolina. He is a deft storyteller, and his transitions are seamless. Whether Bolen is writing about millennia-old cypress, the carnivorous plants of longleaf pine savannas, tundra swans, or blackwater rivers, his writing skips like a pebble on calm water.

In this refreshing read you will move across barrier island beaches and dunes, glide across salt marshes, navigate sounds and coastal rivers, slog through interior wetlands, and search for fox squirrels and the elusive amphibians of coastal uplands. This is a book of discovery and not a book of "lists"—birds, fish, mammals, crustaceans, or wildflowers. You'll delight in the menu of delicious stories and anecdotes regarding great white sharks in Pamlico Sound (an infrequent occurrence), right whales, Outer Banks pheasants, Tar Heel manatees, the "magnificent ramshorn snail," sturgeons, and American shad. I especially enjoyed Bolen's use of geological or topographical features of our coastal zone to explain important ecosystems in our coastal counties. For example, you will learn exquisite details about the formation and age of our Sandhills, the Cape Fear Arch, and Carolina bays. In these and other geological landforms you will be introduced to unique plants and animals—the endangered woodpecker found in the longleaf pine forests of the Sandhills, an unparalleled array of carnivorous plants found on the Cape Fear Arch, and the "islands of biodiversity" in our Carolina bays.

In the world of Eric Bolen and James Parnell everything is connected. This book should be read and reread and then used often as a reference. Bolen and Parnell will help you to ask the right questions. Why are

Venus flytraps found only in a very small patch of North Carolina just a few miles around Wilmington? What events and human actions caused the ivory-billed woodpecker and Carolina parakeet to disappear from the North Carolina landscape?

It is not enough at the end of the second decade of the twenty-first century to write well about natural resources or to have expertise in the climate, geology, flora, and fauna of a region. One must also use such tools to make a difference! Throughout the chapters and in the Afterword, Bolen is a powerful advocate for coastal resources. He will make you uncomfortable as you read about the impact of dams on coastal rivers, the ongoing damage wreaked by invasive plants and insects, and the price of extinction. In the end, this important work is a powerful reminder of our rich coastal heritage, an invitation to explore coastal resources, a primer to help each of us ask the right questions, and the inspiration to be better advocates. In the pages of this gift, I finally experienced the privilege of being in a class taught by Eric Bolen and James Parnell.

Tom Earnhardt, naturalist and writer
and host of UNC-TV's *Exploring North Carolina*
May 2020

Preface

Anthropologist-philosopher Loren Eiseley famously noted that "if there is magic on this planet, it is contained in water." Indeed, water in its various settings stands as the sine qua non of North Carolina's Coastal Plain, whether in the soggy soils of Carolina bays or in the eternal surf of the Outer Banks. To briefly illustrate the breadth of such sites, Lake Waccamaw offers naturalists lessons about the influences of isolation and geology on aquatic fauna, as well as how the Cape Fear Arch supports a flora rich with endemic species. With a bit of imagination, one might envision ivory-billed woodpeckers hammering away in dark cypress swamps bordering the Black River, home of the oldest trees in the eastern United States. In some precious places with poor drainage, pitcher plants digest arthropods while at the same time safely harboring other creatures inside their tubular leaves. On a lucky day afield, one might see—or more likely hear—a Pine Barrens treefrog in one of many pocosins, those dense shrubby bogs where bears may also seek food and cover. And on the Outer Banks, feral horses graze on marsh grasses, paw their own wells for fresh water, and huddle during fierce storms. History and culture also abound on the seaside edges of the Coastal Plain, where the salty water gave rise to down-home carvers of rugged decoys, fostered the woeful chants of strong men hauling in nets heavy with menhaden, and inspired the first of Rachel Carson's eloquent books about coastal marine life.

Within this watery mosaic lie other sites of equal interest. Savannas of longleaf pine, the remainder of a once-vast southern forest, rely on fire for their existence and, when mature, become primary habitat for red-cockaded woodpeckers, the only species of its kind to nest in living trees. In addition to its stately pines, the Carolina Sandhills in places includes sites where turkey oaks and a few herbaceous plants deal with the aridity of the sandy soils. Drier yet, the dunes along the coast support communities that cope with shifting sand, salty air, and hellish storms. Nearby, the dense canopies of maritime forests become temporary havens for migratory songbirds en route to their southern wintering grounds.

There's more, of course, to this magic region of water and sand, but we hope these few sips and bites will tempt your palate into a fuller course of natural history in the pages that follow. Indeed, our title was adopted from the preface of John Lawson's landmark, *A New Voyage to Carolina* (1709), in which he lamented that travelers too often failed to present a useful account of the land and its natural wealth, even "tho' the Country abounds with Curiosities worthy of nice Observation." Lawson thus penned an eighteenth-century view of the natural history of what he famously called a "delicious" land. Now, more than three centuries later, we hope our effort might provide a modern view of the curiosities that abound in what we regard as a particularly delicious region—the Coastal Plain of North Carolina.

The plan of our book is simple—it begins with an overview of the features and biological richness of North Carolina's Coastal Plain and the naturalists they attracted across four centuries of awe and discovery. Thereafter, the text partitions the Coastal Plain into regions, each with its own ecological settings and cast of characters. The chapters deal with beaches, marshes, rivers and sounds, and interior wetlands before ending in the drier uplands. Infoboxes appear throughout the book; these highlight topics that stand alone yet add some flavor to the main text, as, for example, whales and whaling in North Carolina, the invasive abilities of red maple, and the extinction of Carolina parakeets. To conclude, an Afterword outlines a concern shared by all naturalists and, hopefully, many others as well—endangered species. Our world, though already diminished by too many anthropomorphic extinctions, nonetheless remains rich in biological treasures. With ethical and ecological responsibilities in mind, let us strive to keep it that way.

For convenience, the sources we consulted, as well as titles for additional information, are easily accessible at the end of each chapter. Rather than clutter the text, the current Latin names of organisms appear in the appendix, including once-familiar names that are no longer recognized because of insights revealed by molecular taxonomy.

Acknowledgments

"Outstanding" understates the remarkable staff of the UNC Wilmington Library. One and all repeatedly went the extra mile to provide sources and services for our project. Indeed, it's hard to describe their exceptional professionalism; hence, a brief but heartfelt "thank you" will have to suffice to express our indebtedness. Likewise, we appreciate the accommodations afforded to us by the Department of Biology and Marine Biology at UNCW, especially for facilitating our electronic access to the library after we both retired.

Special thanks also go to Tom Earnhardt for writing the foreword, doing so with his deft touch for describing nature and the pressing need to conserve natural resources. Tom, an attorney by profession and a naturalist at heart, serves as producer, writer, and host for an acclaimed TV series, *Exploring North Carolina*. His book, *Crossroads of the Natural World: Exploring North Carolina with Tom Earnhardt*, stands as a "must read" for anyone wishing to learn about the state's varied texture of wild places and natural wonders. So thank you, Tom, for your kind words to introduce our exploration of North Carolina's Coastal Plain.

Our thanks go to the editors—Patrick Allen, Nate Holly, and Jon Davies—at the University of Georgia Press for guiding our work to fruition. We extend our appreciation to the design team at the press for their careful attention to detail. Mary M. Hill's thorough copyedit did much to improve the text. A special shout-out goes to Dan Flores for his sage advice concerning our search for a publisher—"spot on," Dan.

We appreciate the generosity of everyone who provided photos; credit for these appears in the accompanying figure captions. All others were provided by coauthor JFP.

Completion of a book of this scope clearly involves the contributions of many colleagues, old and new alike, to whom we offer our appreciation for so generously sharing their advice and expertise; several also reviewed topics matching their special areas of interest. We alone, however, remain responsible for any errors that might appear in the following pages. We gratefully acknowledge Bill Adams, Lindsay Addison, Jon Altman, James Amoroso, Steven Anderson, Edward Arb, Allyne Askins, Gloria Aversano, Jeff Beane, Joan Berish, Marissa Blackburn, Jesica Blake, Francoise Boardman, Stephanie Borrett, Blair Boyd, Alvin Braswell, Jacqueline Britcher, Jason Brown, L. J. Brubaker, Kerry Brust, Andy Buchanan, Misty Buchanan, Sherrel Bunn, Kemp Burdette, Larry Cahoon, Peter Campbell, Jay Carter, Dale Caveny, Brian Chapman, Sandy Chapman, Cheryl Claassen, Mary Kay Clark, Matt Connolly, Jerry Cook, Margaret Cotrufo, Jan Davidson, Pete DeVita, Alton Dooley Jr., Melonie Doyle, James Ducey, Karen Dugan, Carl Eben, Devon Eulie, Van Evans, Jesse Fischer, Corrin Flora, Dan Flores, D. Wilson Freshwater, Mark Garner, Cyndi Gates, Patricia Gensel, John Gerwin, Matt Godfrey, Walker Golder, James Guldin, Donald Hall, Jeffrey Hall, Katie Hall, Nathan Hall, Toby Hall, John Hammond, Bill Harris, Becky Harrison, Frank Henry, Camilla Herlevich, Paul Hosier, Doug Howell, Stacy Huskins, Christian Kammerer, Jud Kenworthy, Richard Kiltie, Jennifer Koches, Chris Kreh, R. Wilson Laney, Lee Langston, Wanda Lassiter, Richard LeBlond, Gail Lemiec, Lynn Leonard, Dale Lockwood, Michelle Ly, Mike Mallin, Kay W. McCutcheon, Bill McLellan, Sarah McRae, Sharon Meade, Patrick Melvin, James Mickle, Paul Moler, Henry Mushinsky, Paige Najvar, Robert Orth, Ann Pabst, David Padgett, Joe Pawlik, Doug Peacock, Gary Perlmutter, Brad Pickel, Scott Pohlman, Martin Posey, Gloria Putnam, Jean Richter, Stanley Riggs, Spencer Rogers, F. "Fritz" Rohde, Jennings Rose, Chris and Pernille Ruhalter, Tammy Rundle, Gabrielle Sakolsky, Thorpe Sanders, Michael Schafale,

Fred Scharf, Sara Schweitzer, Nathan Shepard, Steven Shepard, Roger Shew, Lora Smith, Clyde Sorenson, Bruce Sorrie, Travis Souther, Dale Suiter, John Stanton, Beth Steelman, John Stoffolano, Susan Stuska, Mary Sword, John Taggart, Beverly Tetterton, Vicky Thayer, Jean-François Therrien, Amanda Thomas, Peter Thomas, Nate Toering, Carm Tomas, Tracey Tuberville, Howard Varnum, Jack Watson, Wm. David Webster, Elisabeth Wheeler, G. Richard Whittecar, Ted Wilgis, Scott Wing, Andy Wood, and Jen Wright.

We remain indebted especially to our wives, Elizabeth and Frances, for their unwavering support of our ventures, past and present.

Eric G. Bolen
James F. Parnell
Wilmington, North Carolina

An Abundance of Curiosities

I

The Coastal Plain of North Carolina and Its Early Naturalists

The Landscape

A journey west to east across North Carolina traverses three regions, each of which is defined using criteria based on geology and topography. A mountainous zone occupies western North Carolina, where ranges and peaks of the Appalachian chain offer majestic views of a terrain unlike any other in the state. Indeed, Mount Mitchell in the Black Mountains is the tallest peak east of the Mississippi River. Most streams arising in these mountains feed rivers whose waters flow westward to the Mississippi River and on to the Gulf of Mexico. A middle zone, the Piedmont, consists of rolling hills and the watersheds for many of the state's larger rivers, all of which flow eastward to the Atlantic Ocean. Finally, our journey reaches the Coastal Plain, about 45 percent of the state and the focus of the following pages.

What is commonly known as the Fall Line but that more accurately constitutes a zone defines the western edge of the Coastal Plain. Here the topography drops from the Piedmont to a relatively featureless plain, all of it less than one thousand feet above sea level and much of it less than five hundred feet in elevation. Along this line, too, the sandy soils replace the hard-rock history of an earlier geological age and instead reflect the origins of a younger, marine-based terrain. Said otherwise, metamorphic rocks of the Piedmont give way to sedimentary rocks of the Coastal Plain. However, the western border becomes even better defined by the tell-tale change in the rivers as they flow from the Piedmont eastward to the Coastal Plain. The relatively rapid drop in elevation produces falls or at least rapids in the river channels—barriers that halted commercial shipping and promoted settlements, then towns and cities, where freight was off-loaded for local processing or for overland transport farther inland. The eastern border, of course, is the coastline, where large lagoons (or "sounds") are protected by an offshore rim of barrier islands that stretches northward from Cape Lookout, whereas to the south, the barrier islands lie closer to the shore, and the sounds are either far smaller or absent.

Rivers coursing through the Coastal Plain typically lack high banks, although those at Neuse State Park offer an interesting exception, and they rarely form distinctive valleys. Instead, river water in the region regularly flows over the low-lying banks into wide floodplains often measured in miles and cloaked in swamps and bottomland hardwood forests. The water characteristically moves "slow and lazy" across the gentle slope of the terrain, and few rapids and riffles occur east of the Fall Line. As they approach the ocean, the rivers encounter salt water, sometimes several miles upstream, and thereby begin their transition into estuaries—species-rich environments of considerable ecological importance yet too often compromised by human carelessness. In estuaries, the lighter fresh water overrides the heavier salt water; hence, two types of biota can occupy the same water column. To better visualize this profile, imagine two wedges, one (fresh water) atop the other (salt water), with the thin ends of each pointed in opposite directions.

Because of its low-lying profile and abundant rainfall, the Coastal Plain represents a land of ecological diversity in the form of wetlands (Chapter 4). Many areas experience long hydroperiods—the length of time the soil remains saturated—both below and above ground and thus discourage many forms of vegetation while favoring others, including some unique communities. Coarse sands drain well, but long hydroperiods result in areas where fine sands and silt dominate the soil profile. Fire also exerts a powerful influence on the communities of the Coastal

Plain in both upland and poorly drained locations. Indeed, some communities, especially those dominated by longleaf pine, persist solely because of the regular occurrence of wildfires. Along the coastline, salinity becomes an ecological force to which relatively few plants have adapted —a circumstance that often results in monocultures of halophytic (salt-loving) species (Chapter 3).

A unique albeit largely hidden geological feature—the Cape Fear Arch—influenced development of a diverse flora in the southeastern region of North Carolina's Coastal Plain. The arch formed when Cretaceous sedimentary deposits were uplifted during the Eocene some forty-five million years ago. The arch extends from Cape Lookout in North Carolina to Cape Romain in South Carolina and reaches inland to the Fayetteville area. Significantly, the arch once formed a peninsula that remained above water when sea levels were higher than at present. This allowed the terrestrial flora to continue evolving while the surrounding areas remained underwater—the isolation of the peninsula favored the evolution of distinctive species. As a result, a significant number of the twenty-two endemic and twenty-two nearly endemic taxa in the Coastal Plain are associated with the Cape Fear Arch. Of these, the Venus flytrap serves as the preeminent example.

Two endemic communities further highlight the region's biodiversity (Fig. 1-1). A wet marl forest developed in a small area of Pender County where a high water table overlies flat-lying deposits of limestone. The combination of seasonal wetness and a calcium-rich soil produces a community that occurs nowhere else on Earth. Nutmeg hickory, a species whose primary range lies in the lower Mississippi River valley, forms a major component of the forest; this disjunct population represents the northernmost occurrence of the species. Roughleaf dogwood, another calcium-loving species from farther south, also flourishes here in one of only two sites in North Carolina. A dense cover of dwarf palmetto beneath the trees stands out as an obvious marker of this community; it provides habitat for a relict population of eastern wood rats. These semiarboreal rodents build bulky houses of sticks and other litter but also collect colorful or shiny objects such as tinfoil and bottle caps—a habit leading to the tag "pack rats." Perhaps surprisingly, while the plants in the wet marl forest thrive in other places, they have not formed a similar community on those rare sites elsewhere where marl soils occur. Regrettably, logging, invasive vegetation, and limestone mining have degraded much of this unique environment, and, lacking protection from a conservation agency, the remainder may likewise face a dim future.

Pine savannas dominated by an herbaceous layer of rush featherling represent the second endemic community; these occur only in Brunswick and Pender Counties. The thick rhizomes of rush featherling, a lily, form hummocks that distinguish these sites from the flat terrain found in other savannas (and make walking difficult). In early fall, the plants bear several white, star-shaped blossoms that arise on the side of an erect stem, a showy display likened by some as "snow in September." Wire grass and Carolina dropseed may codominate at some locations, but the dense hummocks largely exclude other plants; hence, these rare communities lack the richness typical of pine savannas. Rush featherling burns, but it is less flammable than wire grass. This community develops at sites such as the rims of Carolina bays, described in Chapter 5, upland flats, and other locations where coarse, sandy soils become saturated during wet seasons. Rush featherling also occurs in other settings but not as a major component of the local flora.

Overall, the Coastal Plain lacks prominent topographical features, although the attenuated tip of New Hanover County warrants mention. There the mainland tapers to a point bordered to the west by the Cape Fear River and to the east by the Atlantic Ocean at a site known as Fort Fisher, once a stronghold of the Confederacy and now a state park. In effect, the geography functions as a large funnel for migrant birds headed to wintering grounds farther south, some traveling to the Caribbean and South America. Similar funnels but on a larger scale occur at Cape May in New Jersey and at Cape Charles on the Delmarva Peninsula. In the fall, as birds travel south, northwest winds deflect migrants toward the coast, where they then follow the coastline southward, eventually concentrating at the tip of the peninsula—the spout on the geographical funnel. The birds often pause at these locations before continuing their travels, usually to await conditions that favor safe passage across the open water ahead, in this case across the Cape Fear Estuary to the mainland near

FIGURE 1-1. The biodiversity of North Carolina's Coastal Plain includes a small tract of wet marl forest in Pender County, the only community of its type on Earth (*top*) and a few sites where rush featherling dominates the understory of a pine savanna (*bottom*). Photo credits: Michael P. Schafale, North Carolina Natural Heritage Program (*top*) and Robert Thornhill (*bottom*).

Southport. In other words, the birds delay their flight to avoid winds that might otherwise carry them out to sea and away from land. Some birds, in fact, reverse course and briefly fly *north* from Fort Fisher apparently in search of a narrower place upriver to reach the mainland, after which they resume their journey south. These and other aspects of migratory behavior are particularly well documented for birds at Cape May (see the references).

The diversity of North Carolina's Coastal Plain increased in 2005 when wood storks, apparently refugees from states farther south in search of better nesting habitat, began establishing several nesting colonies in what is now their northernmost breeding area in North America (Fig. 1-2). These large, fish-eating birds build their nests in cypress or sometimes gum trees, typically in swamps where alligators limit the access of raccoons and other egg-loving predators. The breeding cycle of wood storks coincides with the normal dry season, which concentrates fish in a few deeper areas and enables the adults to meet the demands of their ever-hungry nestlings. In fact, a family of wood storks—broods vary from one to three chicks—may consume more than 440 pounds of fish during the nesting season. When necessary, the adults will travel forty miles or more to find suitable feeding areas. To capture their prey, wood storks rely on touch, not sight; hence, they can locate fish in murky water. They forage with slightly opened bills, which, on contact with a fish, snap shut with amazing speed—a reflex that takes less than twenty-five milliseconds. Wood storks, federally listed as endangered in 1984, were downlisted to threatened in 2014, thanks to the work of conservation agencies.

The Coastal Plain also wants for physical barriers that might affect the distributions of certain species. The lack of mountains obviously precludes vertical zonation in the region's plant and animal communities. However, the presence of the cold-water Labrador Current streaming southward and the much warmer Gulf Stream traveling northward enables mingling of northern and southern species along the coast. Cape Hatteras, where the Gulf Stream comes closest to land and the water temperature changes abruptly, marks the zone of faunal contact between north temperate and subtropical organisms; the demarcation affects a full range of life, ranging from seaweeds to marine mammals. Among

FIGURE 1-2. Wood storks began nesting in North Carolina in 2005. Wetlands on the Coastal Plain with stands of mature cypress provide the birds with ideal habitat for their colonies.

the latter, long-finned pilot whales and harp seals represent the northern fauna, whereas short-finned pilot whales and West Indian manatees provide examples of the southern fauna. The faunal mixing carries over to terrestrial species as well, and a few, as noted below, reach their limits in the Coastal Plain.

Some examples of species with distributions that extend northward into the Coastal Plain of North Carolina include painted buntings and eastern woodrats; both reach their northern limit along the Atlantic coastline in North Carolina. Similarly, the range of marsh rabbits essentially ends at the northeastern border of North Carolina, although an isolated population may occur just over the line at Hog Island, Virginia. The original range of cabbage palmetto barely extended into North Carolina on Bald Head Island at the mouth of the Cape Fear River. A related species, the cold-hardy dwarf (or bush) palmetto, extends no farther north than a woody island in Currituck Sound. North Carolina also marks the northern limit for eastern diamondback rattlesnakes and American alligators, as well as for pigmy rattlesnakes. Curiously, the coloration of those individuals of the latter species found on or near the Albemarle Peninsula differs from those living elsewhere in the Carolinas. Their bold pattern of red, orange, or pink markings stands out in comparison with the darker, subdued markings of the others.

The southern range of American beachgrass and the northern distribution of sea oats overlap on the Outer Banks. Cape Hatteras more or less represents a midpoint in this ecotone, south of which American beachgrass gradually drops out and sea oats becomes dominant, with the reverse occurring north of the cape. Both species occupy the same key niche in dune systems, where they capture and hold sand in place. A similar exchange occurs in the submersed aquatic vegetation in North Carolina's fabled sounds. Eelgrass, a prominent seagrass farther north, reaches its southern limit at Pamlico Sound, where a southern counterpart, shoalgrass, begins to assume dominance. On the Outer Banks, wetlands at Pea and Bodie Islands mark the southern limit for nesting blue-winged teal, black ducks, and gadwalls along the Atlantic coast. Whereas the teal and black ducks nest continuously in coastal marshes northward into Canada, the gadwalls nesting in the Outer Banks form the southernmost of several isolated colonies that dot the coastline between North Carolina and New York, each of which lies at least sixteen hundred miles from the species' normal breeding range in the interior of North America.

A Wonderland for Naturalists

North Carolina did not go unnoticed by naturalists of the Colonial era.[1] Many recorded their observations in narratives, while others collected

---

1   The Province of Carolina, carved from Virginia in 1663, later split into two colonies that today form the states of North and South Carolina; the date is usually cited as either 1711 or 1712, although some sources claim the separation did not occur until 1729.

FIGURE 1-3. John White, governor of the ill-starred Roanoke Colony, produced numerous paintings of Native Americans and New World creatures, among them a diamondback terrapin in which he inexplicably added a sixth toe to the feet on the right side. Facsimile of White's original watercolor by Elizabeth D. Bolen. Photo credit: Dale Lockwood.

specimens for their own use or for the cabinets of their wealthy patrons. A few were skilled artists who produced colorful portraits of the plants and animals they encountered. But all were pioneers whose skills and curiosity revealed the natural wonders of the New World, not the least of which were those of North Carolina's Coastal Plain.

John White (ca. 1540–1593) stands as the earliest of the naturalists to visit what is today North Carolina (then part of Virginia). His vivid watercolors included crabs, fishes, butterflies, and sea turtles, as well as a visual record of Native Americans and their way of life. His work included considerable detail, in one case showing the tail of a puma that adorned the clothing of a chieftain. White sketched his subjects in graphite pencil before adding washes made from a variety of colorful pigments (Fig. 1-3). The paintings were published with a text written by Thomas Harriot (ca. 1560–1621), a scientist who accompanied White on a voyage to "Virginia" in 1585; both were members of an expedition charged with founding a colony on Roanoke Island. The effort failed, however, and the colonists returned to England the following year. But more was to come, as told in a story well stitched into the fabric of North Carolina's history.

In an expedition sponsored by Sir Walter Raleigh (ca. 1554–1618), White returned to Roanoke Island in 1587, this time as the governor of a community of more than one hundred settlers who sailed with him. Times were hard, however, and their meager supplies quickly diminished, circumstances worsened by the failure of their crops.[2] White thus went back to England, planning to resupply the struggling community, but, deterred by the threat of Spanish warships, he was delayed until 1590. On his return, he found the site abandoned, and the settlement became known to history as the fabled "Lost Colony." The mysterious fate of the settlers included his own daughter and son-in-law and their daughter, Virginia Dare, the first English child born in North America. Still, White's paintings live on as beacons into a world untouched by European influences, sadly including an ancient culture now forever vanished. Similarly, some of his artwork—a loggerhead turtle, for example—represents species today listed as threatened, but his paintings also included an eastern tiger swallowtail, a species formally recognized in 2012 as North Carolina's state butterfly. More than a century would pass before another naturalist, Mark Catesby, would again illustrate the natural history of the Carolinas (see below).

In the spring of 1700, London-born John Lawson (1674–1711), while in search of places to travel, chanced on a gentleman who "assur'd me, that Carolina was the best Country I could go to; and, that there then lay a Ship in the Thames, in which I might have my passage. I laid hold of this Opportunity." Lawson immediately set sail on a voyage that ended in present-day Charleston, then a nine-acre "metropolis of South Carolina." There, in due course, the lords proprietors engaged Lawson to reconnoiter the backcountry, an unspoiled and ill-defined territory then known only to a few Indian traders and, much earlier, to gold-seeking Spaniards. Thus began, almost whimsically, one of the great adventures in the tableau of natural history into a land whose wonders were at the time virtually unknown to Western science.

Lawson and a few companions at first paddled a large dugout canoe northward along the coast, where they eventually reached the mouth of the Santee River and followed the river inland. They later abandoned the canoe and trekked overland, entering North Carolina near Waxhaw in

2   See "Ancient Trees on the Black River," Chapter 4.

mid-January 1701. From there, Lawson and his party continued north until they reached the vicinity of modern Raleigh-Durham and then turned east to complete a huge arc that ended near present-day Washington, North Carolina. Lawson reckoned he had traveled a "thousand miles" on his fifty-seven-day trip, but in reality, the route covered about 550 miles, not counting twists and turns. Lawson, "pleas'd with the goodness of the country," established Bath in 1706, the state's first incorporated town, although he settled along a stream, still known as Lawson's Creek, near what is now New Bern. During this period, he served as the court clerk and registrar for the county; in 1709, as the official surveyor for the lords proprietors, he published a map of North Carolina.[3]

Lawson kept a journal during his 1700–1701 journey that, supplemented with additional observations, he steadily worked into a manuscript suitable for publication. To further that end, he returned to England in 1709 and quickly found a willing publisher in John Stevens, a prominent scholar who issued Lawson's work serially in an anthology about world travel and attendant discoveries. Later that year, when the series was completed, Stevens compiled the installments into a single volume entitled—in its short form—*A New Voyage to Carolina; Containing the Exact Description and Natural History of That County.*[4] His task completed, Lawson returned to North Carolina in 1710 on a mission that produced a Swiss settlement known as New Bern and the state's second-oldest community. Tragically, rebellious Tuscaroras captured and killed Lawson the following year as he explored the upper reaches of the Neuse River. Ironically, he apparently died from one of the torturous practices that he had described in his book, in which a prisoner was first impaled with pitch-coated pine splinters, thus resembling a human porcupine, which were then ignited to "burn like so many Torches; and in this manner, [forced] to dance around a great Fire . . . till he expires."

3   The map extends from the Atlantic coastline to the Piedmont of both North and South Carolina—the Coastal Plain—and essentially the only area falling within the vague boundaries of geographical knowledge at the time. Lawson's map presents a surprisingly good representation of the Outer Banks and river systems, given the instruments and techniques available to surveyors of the Colonial era.

4   In 1737, John Brickell published *The Natural History of North Carolina*, which drew heavily from—some say plagiarized—Lawson's book.

A New Voyage, much of which deals with North Carolina, offered a trove of new information and even foretold a problem arising far into the future: bald eagles dying from lead poisoning. Lawson wrote, "The bald Eagle attends the gunner in winter . . . and when he shoots and kills any fowl, the Eagle surely comes in for his Bird [and] those that are wounded, and escape the Fowler, fall to the Eagle's share." In other words, Lawson observed that bald eagles commonly fed on wounded ducks and thereby ingested shot embedded in the carcasses, which we now know results in the fatal effects of secondary lead poisoning. Indeed, centuries later these circumstances contributed to a ban on lead shot for waterfowl hunting. Lawson also recorded that "white Brant" (snow geese) flocked to newly burned coastal marshes and savannas, which they "tear up like hogs" in search of the "roots of Sedge and Grass," an account that may represent the first published example of fire ecology. He suggested burning as a way of attracting snow geese, which in modern times became a management practice to improve the availability of their winter food resources. More recently, however, snow geese have become so numerous in some areas that they denude large areas of coastal marshes with ruinous "eat outs," some of which may be severe enough to hinder revegetation of the site.

Lawson developed a healthy respect for alligators, especially after one built a nest under his house at New Bern. One night the "roar" of this "ill-favour'd neighbor . . . made the House shake about my Ears," leading him to think the ruckus might be a ruse initiated by Indians. Chagrined when he learned the truth of the matter, he nonetheless included this story on himself in his book, which suggests a humorous side to his personality. As a more practical matter, Lawson also noted how the teeth of dead alligators served as "chargers" (ramrod tips) for loading guns, and because the teeth were of various sizes, they were "fit for all loads." On visiting the coastline, presumably the Outer Banks, Lawson commented that bluefish in "their eager Pursuit of the small Fish" sometimes "run themselves ashoar [where] they cannot recover the Water again." He also observed that the fish arrive in the "fall of the Year . . . when there appear great Shoals of them." On such occasions, the Hatteras Indians waded into the shallows to strike the fish, just as today's anglers await these runs with hook and line. The beaches, he noted, provided nesting sites for three species of sea turtles, with green sea turtles the least common and hawkbills more so; both provided "extraordinary Meat." Loggerheads, according to Lawson, did not make good table fare, although the eggs of all three species were "very good Food." Today only the loggerhead regularly nests along the Outer Banks—the northernmost extent of its range on the eastern coast of North America—where it is rigorously protected as an endangered species (as, sadly, are all species of sea turtles). Much more might be added, of course, but to conclude our limited review of Lawson's important work, we cite a line taken from the preface of *A New Voyage* in which he speaks of America as a country that "abounds with Curiosities worthy of nice Observation" and to this day has served as a guiding light for others.

Mark Catesby (1682–1749), a well-to-do Englishman, made two trips to Colonial America; the first of these began in 1712 with a seven-year visit to Virginia, the second in 1722 for three years in South Carolina—locations that bracket present-day North Carolina. He traveled extensively during both of these visits, including trips inland to the Piedmont and to Florida and the Bahamas, collecting plants and seeds for English patrons who fancied "curiosities" from the New World, as well as snakes and other small animals. He preserved the latter in rum-filled vessels, which at times were drained during shipment by thirsty sailors who clearly had little concern for hygiene, let alone zoology. Before leaving on his second voyage, Catesby had agreed to take over a project—a full treatment of America's natural history—that had been assigned earlier to John Lawson but that ended with his untimely death. Hence Catesby's second visit, sponsored by certain members of the Royal Society of London, was dedicated not only to collecting additional specimens but also to recording and illustrating objects of interest for a wide-ranging treatise unlike any other of its day—the first illustrated flora and fauna of early America.

Catesby returned to England in 1726 to begin working on his book, *Natural History of Carolina, Florida, and the Bahama Islands*, which came out in two volumes after some two decades of work; the first volume appeared in 1732, the second in 1743, followed by a supplement in 1747. When completed, the work included illustrations of 109 birds, 33 amphibians and reptiles, 46 fishes, 31 insects, 9 mammals, and 171 plants, which were supplemented with text describing the geography, soils, and climate of each region. He also presented a section devoted to Native Americans, along with an acknowledgment to Lawson as a source for much of that material. Catesby also chose a format that followed Lawson's book, at times drawing from his species accounts as well, and indeed he copied some of the illustrations John White produced in an earlier century. In keeping with the taxonomic style of the day, Catesby described the species he studied with polynomials, but these were no longer accepted when a binomial system became a standard that remains in place today. As a result, he lost credit for his discoveries of new species when the scientific names in his book were later replaced with binomials.

Catesby's *Natural History* remained the most authoritative work on birds until Alexander Wilson published his *American Ornithology* early in the nineteenth century (see below). Thomas Jefferson consulted the book, as did Meriwether Lewis and William Clark and a host of naturalists on both sides of the Atlantic. Catesby set the record straight on a few prominent myths, among them that, when snapped, the tail of a coachwhip snake could cut a man in half. He also missed the mark on occasion, as when he reported that the guts of the now-extinct Carolina parakeet are "certain and speedy poison" to cats. Nonetheless, Catesby's paintings were the first to show birds and other animals in association with vegetation, usually plants of some ecological importance, such as a food source or a nesting habitat. His illustration of bobolinks, for example, includes a stalk of rice in full fruit, which, in season, becomes their food of choice and made them a scourge for rice-growing plantations along the Carolina coastline. Catesby also depicted animals feeding or otherwise actively engaged, as opposed to the stiffly posed flat portraits typical of the era's artwork. He illustrated a cardinal, the state bird of North Carolina, perched on the branch of a hickory. Not until Audubon would these artistic features again appear in wildlife illustrations. Catesby also found time to present his views on the migratory behavior of birds to the Royal Society of London, to which he was elected a fellow in 1733. In particular, he refuted the idea that birds hibernated in winter, supposedly in the muddy bottom of ponds! He explained that birds in fact disappeared because of their seasonal travels to other areas.

A New Taxonomy

A milestone in natural history occurred during America's Colonial period when Swedish physician and botanist Carl Linnaeus (1707–1778) devised an elegant yet simple way to formally identify species—the binomial system. Instead of a long string of Latin nouns and adjectives, which often varied considerably in the description for the same organism, the new method relied on just two words, the first being a genus (noun) followed by an epithet unique to the species (adjective). Hence, *Typha latifolia* today identifies broadleaf cattail, whereas *Typha angustifolia* denotes a related species, narrowleaf cattail; such combinations remain the universal means for identifying these species apart from all others. Epithets such as *latifolia* may be used again to describe a species in another genus that also has broad leaves.

Not surprisingly, many species are identified with *carolinensis* (e.g., the eastern gray squirrel [*Sciurus carolinensis*]), an epithet rivaled only by *pennsylvanicus* and *virginianus*, two other colonies/states closely associated with the early days of natural history. Moreover, each species assumed its place in a hierarchal series of groupings (taxa)—genus, family, order, class, phylum, and kingdom—based on characteristics shared in common at each level. In 1758, Linnaeus presented his system, for both plants and animals, in the tenth edition of *Systema Naturae*, which established him as the "father" of classification. But for Catesby and others, the new system disenfranchised their formal recognition as discoverers of new species simply as the result of Linnaeus replacing the original nomenclature with binomials. In this manner, Linnaeus gained formal credit for discovering "new" species that were in fact collected by other naturalists.

Modern techniques using molecular genetics—DNA analyses—have sometimes revealed heretofore "hidden" taxonomic relationships. These, in turn, have occasionally resulted in replacing several long-established Latin names familiar to naturalists. As an example, *Sporobolus* has replaced *Spartina* as the generic name for both smooth cordgrass and marsh hay. The currently accepted Latin names for species mentioned in this book appear in the appendix.

Catesby is also remembered today in the names of several species, among them Catesby's lily (*Lilium catesbaei*) and the most familiar of all, the American bullfrog (*Rana catesbeiana*).

King George III's interests in Colonial America embraced more than stamp taxes, naval stores, and alliances (or wars) with Indians. Indeed, the flora of the New World had captivated the imagination of Europeans much as its fauna had with marvels such as a snake that rattled its tail and a pouched creature that played dead to escape trouble. So in 1765 the king appointed a Philadelphia Quaker named John Bartram (1699–1777) as His Majesty's botanist for North America. Bartram lacked formal training in science but nonetheless learned to identify and propagate plants far better than anyone else in the colonies. His charge was to find and collect "natural productions" primarily in East and West Florida, newly ceded to England by Spain in exchange for Cuba. But the rich flora of North Carolina also played a role in his celebrated career as a naturalist and later that of his son William (1739–1823) as well. The family ties with North Carolina began in 1709, when John's father and wife moved to land he bought on Bogue Sound near present-day Swansboro, although John remained in Philadelphia. In 1711, the same uprising by Tuscarora Indians that killed John Lawson also claimed the life of John's father; his wife and children, although captured, were later rescued and returned to Philadelphia, where John, despite his youth, was managing the family farm. In the fall, with the harvest completed, John ventured afield in what was still untamed wildness in search of new plants and seeds to collect.

About 1728 John built a sturdy stone house on the west bank of the Schuylkill River that became a meeting place for those with scientific interests, among them George Washington, Thomas Jefferson (who bought plants for Monticello), Benjamin Franklin, Alexander Wilson (see below), and fellow botanists such as Frederick Pursh and Thomas Nuttall. Other visitors included Tar Heels such as Alexander Martin, who later served several terms as the state's governor, and Hugh Williamson, a student of snakes, comets, and climate. Both men were elected members of the American Philosophical Society, the nation's oldest scientific society. John's house stood, and still stands, amid gardens where the botanist grew many of the plants he discovered, one of which is known as Franklin's "lost

FIGURE 1-4. The home of John Bartram, built near Philadelphia in 1728, stands next to his now famous garden and arboretum, which includes plants he collected in southeastern North Carolina and elsewhere, at times accompanied by his son, William. A bloom of Franklin's "lost tree" appears in the foreground. Photo courtesy of the John Bartram Association / Bartram's Garden.

tree," a species no longer known in the wild (Fig. 1-4). To sate their fascination with American flora, wealthy Europeans of the day commissioned Bartram to acquire seeds to plant on their estates, which provided the botanist with additional income; as royal botanist he received the modest sum of fifty pounds per year. Others desired specimens for their "cabinets," the then-current term for private collections of minerals and biological items collectively known as curiosities. Today, Bartram's Garden and Arboretum stands as the oldest surviving garden in the United States. Open to the public, the site is a National Historic Landmark.

Meanwhile, John's half-brother William, a childhood captive of the Tuscaroras, returned to North Carolina, where he served as a colonel in the militia and several terms in the legislature. He prospered at Ashwood, his estate that overlooked the Cape Fear River in present-day Bladen County. Ashwood became a convenient and hospitable rest stop during John's botanical travels. But in John's words, North Carolina was itself also "a ravishing place for a curious botanist," and he delighted in examining species he had never seen, among them a dwarf palm and a bay tree. Although plants remained his central focus, John also took stock of other features of the area's natural history on his daily trips from Ashwood. In one case, the presence of petrified wood protruding from the banks of Willis Creek in southwestern Cumberland County caught his attention; in another, he noted the aquatic life in Lake "Wocoma," his spelling for Waccamaw, including "fish of several kinds" that would later be confirmed as occurring nowhere else.[5] He also noticed the prominent limestone formation at the lake, whose ecological significance would emerge centuries later (see Chapter 5).

In 1761, John sent his son William to Ashwood, where the young man was to establish a trading store, but the business was short-lived, as were his father's hopes to receive more seeds from the area. In 1765, when John began his long journey to Florida, he again stopped at Ashwood, this time to turn his would-be merchant son into his traveling companion and field assistant. Thus, father and son together continued to Florida. It was on this journey that they encountered Franklin's lost tree at a site on the banks of the Altamaha River in Georgia. John returned to Philadelphia the following year, while William remained to try his hand at growing rice and indigo. Again, he failed in this endeavor and rejoined his father in Philadelphia. At this point, William started his own journey (1773–1777) through the southeastern colonies in search of natural productions. The result was a book still celebrated as a premier work of natural history, *The Travels of William Bartram through North & South Carolina, Georgia,*

---

5   Paleobotanist James Mickle identified a specimen EGB collected at the site as a conifer in the genus *Agathoxylon*. It dates to the Santonian Age, which prevailed some 86 to 83 million years ago in the Upper (Late) Cretaceous Period. Because of the considerable difficulties to distinguish taxonomic features in petrified wood, identifications of species within *Agathoxylon* and other genera remain rather "iffy," as witnessed by the plethora of attempts to do so (see Philippe 2011). We extend our appreciation to Dr. Mickle for sharing his expertise with us.

East & West Florida, or, more simply, *Bartram's Travels*, which also presented valuable insights about the Indian cultures he encountered.

William Bartram's expertise included ornithology, and, indeed, Elliott Coues (see below) regarded him as the "god father" of that discipline. During his journey he encountered a colorful vulture whose description almost perfectly fits that of a king vulture, yet the species has never again been seen in the United States—today one must visit the tropical regions of Latin America to find these showy birds. William was also a skilled artist and produced numerous watercolors of plants that are still regarded as among the best of their kind. With a degree of poetic providence, William died while tending the family's beloved gardens. In all, the father-son team stands at the pinnacle of early botanists in North America—John for his luxuriant gardens and conservation of rare plants, William for his monumental *Travels*.

Ornithology made a significant advance when Alexander Wilson (1766–1813) began collecting and painting the birds of eastern North America. Scottish born and a peddler with a bent for poetry, Wilson ran into trouble when he published "The Shark," a poem critical of a wealthy mill owner's treatment of his workers. Legal proceedings followed based on the claim that Wilson had demanded "hush money" to suppress the offending poem. In 1794, after two years of litigation, Wilson realized the seriousness of his predicament and sailed for America. Settling in the Philadelphia area, he briefly worked as a weaver and peddler before turning to teaching school. In the course of these events, he roamed the woods near Bartram's gardens, taking note of the flowers and trees, as well as of the birds attracted by the varied vegetation. Wilson also happened by chance to meet Charles Willson Peale and his son Rubens, who, after Wilson scolded them for shooting a cardinal, explained the usefulness of scientific collections.[6] More important, they arranged for Wilson to visit William Bartram, whose influence shaped the future

6   Peale (1741–1827), a renowned portraitist, also established a prominent museum of "curiosities" in Philadelphia, and the encounter with Wilson occurred on a field trip to procure specimens for his exhibits. He and his sons, several named for Old Masters such as Rubens, Rembrandt, and Raphael, played significant roles in furthering natural history in the early years of the country.

"father" of American ornithology. Wilson began collecting and drawing birds in 1803 and soon planned to publish an exhaustive treatment that covered much of the new nation. To do so, he would need birdshot and black powder as much as brushes and paint to complete his work (just as did Audubon two decades later). Because these naturalists lacked cameras to freeze an image, they used shotguns to secure the models that, once felled, could be arranged in more-or-less lifelike poses.

Wilson completed his great work, *American Ornithology*, in installments, selling them to subscribers—among them, Thomas Jefferson—as a way to fund his travels to find new birds to illustrate. One such trip, in 1809, included North Carolina, where he found travel quite rigorous, especially where wetlands interfered with his progress. Wilson sold subscriptions in Washington, New Bern, and Wilmington and added several species, among them the fox sparrow, black vulture, and yellow-rumped warbler, to his growing folio of paintings. He also believed he had discovered a species new to science, the red-cockaded woodpecker, which, unknown to him, had been described years earlier by a French ornithologist, Jean Pierre Vieillot. But it was Wilson's encounter with another woodpecker, this one of exceptional size, that highlighted his southern tour.

Traveling alone, Wilson neared Wilmington along a deeply rutted roadway that followed what is now the well-paved U.S. Route 17. Much of the way traversed forests of longleaf pine, but it also passed dark cypress swamps, where he "found multitudes of birds that never winter with us in Pennsylvania, living in abundance." About twelve miles north of Wilmington near present-day Scotts Hill, Wilson encountered what he described as a "majestic and formidable species in strength and magnitude [that] stands at the head of the whole class of woodpeckers hitherto discovered [and] the king or chief of his tribe." The painter-naturalist had chanced upon an ivory-billed woodpecker.

At the time, ivorybills were not uncommon in the mature swamp forests of North Carolina, well before these and other pristine environments were lost to the steady march of settlement. Thus, with little effort, Wilson shot three of the birds from trees near the roadway, killing two outright but only slightly wounding the wing of another. The injured bird became his traveling companion for the remainder of the journey to Wilmington, no doubt to serve as a living model. Although Wilson covered the bird, it uttered a "loudly reiterated and most piteous note, exactly resembling the violent crying of a young child." In town the bird's screeches quickly attracted a wealth of concerned citizens, not the least of which were women "with looks of alarm and anxiety." Wilson, in fact, was enjoying himself at the expense of the distressed townsfolk, including hotel proprietor William Dick, when Wilson asked for a room for himself and "my baby." After savoring the moment a bit longer, Wilson uncovered the bird for all to see, which produced a "general laugh," largely aimed at the indignant Mr. Dick. But more was ahead, again likely very much to the displeasure of the hotel owner.

Duly registered, Wilson left the bird unattended in his hotel room while he sought care for his horse. As a very large species of woodpecker, the bird quite naturally also had a very large bill, and a wall in the room offered little resistance to an attack by this avian jackhammer. When Wilson returned, he found a fist-sized hole through the plaster and lathing to the weatherboards, just shy of opening outside and becoming an escape route for the determined woodpecker. Undaunted and still needing the bird for his painting, Wilson tethered the ivorybill to a table in his battered room and left again, this time in search of food for the destructive bird. On returning a second time, Wilson heard the bird hard at work once more, but instead of a damaged wall, Wilson discovered an "almost entirely ruined mahogany table to which [the bird] was tethered and on which he had wreaked his whole vengeance." Wilson did not record Dick's response to either of these events, although it likely featured some colorful language. Nonetheless, the harried innkeeper bought a subscription from Wilson, perhaps at a hefty discount, a notion proposed by historians Marcus Simpson and Donald McAllister in their full account of Wilson's southern tour.

Unfortunately, the wounded bird died a few days later, probably because it rejected handouts from Wilson, but it lives on in a stunning color plate appearing in *American Ornithology*. As a further memorial, Wilson's ivory-billed woodpecker stands as the only accepted record of the species in North Carolina, and the location at Scotts Hill remains as the northernmost point in the known range of the birds on the East

Coast. Both records, however, may better reflect the distribution of ornithologists at the time than they do the former range and abundance of the birds in the state. Some suggest, for example, that ivorybills likely inhabited the bottomland hardwood forests in the Waccamaw River watershed in North Carolina.

Ivory-billed woodpeckers eventually vanished from North America, the victim of destroyed habitat—large tracts of mature forest, including ample numbers of dead and decaying trees as sources of insect larvae. Occasional claims of sightings raise some hopes but more often produce controversy. Such hopes are often dashed when the "ivorybills" turn out to be pileated woodpeckers, a large and relatively common species found widely in the forests of eastern North America. The last accepted records for ivorybills in the United States occurred in Louisiana during the 1940s. A subspecies may persist in Cuba, although the last confirmed sighting occurred in 1986.

Wilson continued from Wilmington to South Carolina and Georgia, eventually gaining enough additional subscribers to assure continuation of his work. In March 1809 he sailed from Savannah to New York City, continuing by stage to Philadelphia. A year later, Wilson journeyed down the Ohio River on a similar mission, highlighted by his chance encounter with Audubon, who had yet to produce his own famous folio of bird paintings.[7] Audubon, in what might have been more than a touch of irony, almost subscribed to *American Ornithology*. Wilson died unexpectedly in 1813 without again visiting North Carolina.

Officers serving as surgeons in the army during the nineteenth century often "practiced" both natural history and medicine. From these ranks, Elliott Coues (1842–1899) emerged as a great naturalist who helped establish ornithology as a legitimate branch of zoology. In 1861, Coues (pronounced "cows") earned a degree from what is now George Washington University; a year later he was appointed a medical cadet

---

7   Audubon, in the 1831 installment of his monumental *Birds of America*, plagiarized Wilson's illustration of a Mississippi kite, which is generally considered one of Wilson's best. The expanding breeding range of Mississippi kites now includes North Carolina, primarily in the bottomland forests along the Roanoke and Cape Fear Rivers.

and called to duty in the Civil War. Coues received an MD from the same university in 1863, which also awarded him a PhD in 1869. Like other naturalists of the era, he received no special training in ornithology.

At age nineteen Coues published a monograph on sandpipers, which initiated a list of papers that eventually numbered in the hundreds—of which more than five hundred titles concerned birds—and perhaps reaching a thousand if a complete count were possible. Much of his output was closely linked to his military postings, the first of which was Fort Whipple in Arizona, where he studied both desert reptiles and birds. On one occasion, thirsty troopers drained a keg of alcohol, no doubt as fortitude for fighting Apaches, and belatedly found that it also contained pickled snakes and lizards. Coues dryly noted that a few of the men looked "decidedly pale about the gills" after discovering the specimens. In 1869, Coues was deployed to Fort Macon, located on the eastern tip of Bogue Banks, which he regarded as "a good place for natural history" because of the diversity of environments close to the fort.[8] However, while the area was free of hostile Apaches, Coues nonetheless thought it lacking in comfort because of ticks, mosquitoes, and malaria—the festoons of Spanish moss in the swamps just across Bogue Sound were in his words "an infallible index" to the threat of disease, something of an ecological approach to epidemiology.

Coues enlisted the services of his hospital steward, whom he trained to prepare many of the birds he collected. A shotgun was the primary tool of ornithologists of the day, and Coues kept his aide quite busy with needle and thread. In fact, the major who commanded Fort Macon, like those wherever Coues was posted, wondered if he paid more attention to the local fauna than to his patients. In addition to his ornithological studies at Fort Macon, Coues turned his attention to marsh rabbits (he called them "hares"), which he regarded as the "most abundant and characteristic mammal on the island." The resulting publication described

---

8   Early in the Civil War, rebel troops commandeered the five-sided fort, which fell to Union forces in 1862 after a relentless bombardment and remained in Union hands for the rest of the war. Fort Macon, a masterpiece of brick construction completed in 1834, became a popular state park in 1936. A 3.2-mile nature trail, named for Elliott Coues, allows visitors to experience a maritime forest, dunes, and salt marshes.

not only the rabbits, their distribution, and their habitat but also their gait, their differences from cottontails and other species (including western jackrabbits, which he surely knew from Fort Whipple), and their external parasites. He also drew upon his medical training and added the anatomical details of the rabbits' skull structure. Coues also studied reptiles and insects while stationed at Fort Macon, but birds remained paramount, and he indeed might well be dubbed the "Bird Man of Fort Macon" for his studies at that location.

Horseshoes crabs, the modern survivors of an ancient group related to spiders and scorpions instead of true crabs, were among the marine life Coues observed at Fort Macon. Each spring, legions of these odd invertebrates spawn at the water's edge, where the females, each one attended by one or more smaller males, lay up to eighty thousand eggs. The event marks one of nature's extraordinary synchronies—the crabs breed just as migrating shorebirds arrive. Hence the abundant eggs offer the famished birds a wholesome meal to fuel the next leg on their long journey to nesting areas in the Far North; the feast enables some to double their body weight in two weeks. Ruddy turnstones, which Coues noted as "very common during migration," are among the shorebirds that depend on the rich bounty of eggs, and at least one major breeding area for horseshoe crabs on the Atlantic coast is now protected expressly for that reason (Fig. 1-5).

Coues marveled at the "incalculable" numbers of waterfowl that overwinter along the coast of North Carolina, where the birds find a "congenial latitude [and an] abundant supply of food." But he also recognized how market hunters subjected the birds to "incessant and systematic persecution" along the coast, but especially at Currituck Sound. Geese, he noted, sold for a dollar or less per bird and "attested to their abundance" near Fort Macon. As described in Chapter 4, market hunting persisted unchecked until 1918, when the Migratory Bird Treaty Act finally outlawed the wanton killing of ducks, geese, and even shorebirds as a commercial venture. Coues capped his studies in North Carolina with a five-part treatise entitled *Notes on the Natural History of Fort Macon, N.C. and Vicinity*, parts of which were coauthored with Henry C. Yarrow, who assumed Coues's position as surgeon when the latter was reassigned to Fort McHenry. The work covered birds, mammals, and fishes, as well as invertebrates.

FIGURE 1-5. While stationed at Fort Macon, army surgeon and naturalist Elliott Coues observed ruddy turnstones gorging on horseshoe crab eggs, which fueled the birds' long migration northward. Photo credit: Eric Reuter

But Coues forever established his prominence as an ornithologist when he published a monumental work entitled *Key to North American Birds* in 1872 that subsequently appeared in five more editions. The account considered every species of living and fossil birds known in the United States and Canada; the introduction alone exceeded sixty pages, in which terms and taxonomic principles were explained to readers at every level. During the 1870s, Coues also played a leading role in the "Sparrow War," in which he vigorously opposed the introduction of house sparrows in North America, whereas another prominent ornithologist, Thomas Mayo Brewer, with equal vigor held an opposing view. At issue was whether the sparrows controlled insect pests; Brewer believed they did and thereby justified their introduction from Europe. Actually, the birds were already at large in the United States, having been initially released in New York City in 1851, but Coues wanted them destroyed by any means possible. Lacking any natural control, the sparrow population expanded across the continent, reaching California in 1910, but on the way they outcompeted bluebirds, purple martins, and tree swallows for nesting cavities. House sparrows currently occur from the Northwest Territories of Canada to southern Panama, where they are widely regarded as pests.

Coues continued studying birds at other postings before leaving the army in 1881 to accept a full-time position as naturalist and secretary for the U.S. Geological and Geographical Survey. He also found time to write volumes about expeditions of Lewis and Clark and other western explorers. Curiously, Coues also developed an interest in a spiritual movement known as theosophy, about which he wrote books and pamphlets, but he later separated himself from the movement. He was a founder and later president of the American Ornithologists' Union. The AOU annually presents the prestigious Coues Award for scholarly work having an impact on ornithology in the Western Hemisphere. At age thirty-four he became the youngest member of the National Academy of Sciences. Few have matched the accomplishments of Elliott Coues, and for visitors to the ramparts of Fort Macon, the lingering presence of this great naturalist can still be sensed in the mews of gulls as they wheel above the grassy dunes.

Into the Twentieth Century

The name Brimley stands alone in the annals of North Carolina's natural history—twice over, in fact. Indeed, Herbert Hutchinson Brimley (1861–1946) and Clement Samuel Brimley (1863–1946) might be justly tagged as "the founding brothers" of a new era when each, singly and jointly, introduced science to the study of the state's rich treasure of biological resources.

Their story begins not in the wilds of North Carolina's craggy mountains or dark coastal swamps but in the English countryside, where the brothers were born into a family of farmers. When not busy with their agrarian chores, the boys collected birds' eggs and "meddled with living things in general," according to John E. Cooper, activities that presaged their future. Times were not good for crops, however, and the family decided to relocate in either Australia or Canada. Then, by sheer chance, the family encountered an official of the recently established North

Carolina Department of Agriculture, Immigration, and Statistics—a marvel of bureaucratic invention. Impressed by what they heard, the family immigrated to North Carolina, ending up in Raleigh in 1880. HH at first tried farming and then turned to teaching, but both efforts proved unsuccessful. In the former, he failed to cope with the Piedmont's rocky soils; in the latter, his English accent vis-à-vis the southern drawl of the students precluded effective communication (or at least that's how HH liked to explain it).

Sometime around 1883, a fifty-cent book, *Taxidermy Without a Teacher*, became the catalyst for an enterprise known as Brimley Brothers, Collectors and Preparers. The careers of the young men thus wheeled into a new direction based on natural history. The market for the frogs, turtles, and other biological supplies they collected included educational institutions, but the mounts of large animals were sought primarily by well-to-do private collectors, who often vied with one another over the magnitude and content of their respective menageries. The Brimleys' knowledge of North Carolina's fauna, till then largely unstudied, expanded rapidly, and the brothers were soon recognized as the South's leading naturalists (Fig. 1-6). Because of his skills as a taxidermist, HH was commissioned by the Department of Agriculture to prepare an exhibit of waterfowl and fishes for display at the State Centennial Exposition (1884). For his salary, HH was carried on the books as a "fertilizer inspector." With the same job description, he later (1893) produced North Carolina's zoological exhibit at the World's Columbian Exposition, often known as the Chicago World's Fair. Among his many writings, which included poetry, HH coauthored *Fishes of North Carolina* and *Birds of North Carolina*, both still standard works.

A particularly significant event also developed in 1893: the Department of Agriculture expanded its museum, which included space where HH could permanently display his work along with the department's own exhibits. Two years later he was named curator of the small museum, which eventually moved into its own and far grander facilities. In 1928, HH became its director, retiring from that position in 1937; nonetheless, he continued to serve as curator of zoology until his death in 1946. Thus, for half a century HH had nurtured an institution that steadily grew from

FIGURE 1-6. C. S. (*left*) and H. H. Brimley (*right*) pose in their taxidermy workshop at what was then the North Carolina Museum of Natural History, later renamed the North Carolina Museum of Natural Sciences. Photo courtesy of the North Carolina State Archives.

a humble beginning into one of the best-known museums of its kind in the United States. Today the North Carolina Museum of Natural Sciences includes two multistory buildings, the Nature Exploration Center and the Nature Research Center, both located in downtown Raleigh. Three satellite facilities have been established elsewhere. It is the largest museum in the southeastern United States and enjoyed each year by more than one million visitors. In 1941, the museum designated its library as the Brimley Library of Natural History in recognition of HH's decades of service; following his death in 1946, it was renamed the H. H. Brimley Memorial Library and features a collection of about seventeen thousand volumes and access to a full range of journals in support of the museum's mission.

No account of HH's long career can fail to mention a fifty-four-foot-long sperm whale that washed ashore at Wrightsville Beach in 1928 (Fig. 1-7). HH wanted the whale's skeleton, but the task would prove far more challenging than he anticipated—the carcass weighed an estimated fifty-five tons, smelled in his words like a skunk factory, and ultimately required two years to process. Before all was over, the whale had been aptly dubbed Trouble. Initially, Harry Davis, who later became the museum's second director, removed the leviathan's jaw, a highly prized part of all vertebrate skeletons, and placed it in the care of the local police chief. Because of the stench, town officials wanted the carcass towed twenty miles out to sea, but HH instead proposed towing it about the same distance to an isolated site on Topsail Island, where it could be buried and later exhumed. The alternate route was acceptable, but it took two days of hard work to budge the carcass from the beach. By then the whale had been ashore for nine days and despite the odor had attracted some fifty thousand sightseers.

The "trouble" began when the carcass broke loose from its mooring on Topsail Island and drifted to a shoal about half a mile out to sea. Now it would have to be cut up at low tide from a nine-foot-high platform and the pieces ferried ashore in small boats. A slit discovered in the whale's back during the process suggested that the thrust of a whaler's lance might be the cause of death. The troubles continued when the Topsail residents complained that the discarded flesh was ruining fishing, although HH countered that it more likely attracted greater numbers

FIGURE 1-7. Thanks to a two-year effort of H. H. Brimley, the skeleton of Trouble, a large sperm whale beached at Wrightsville Beach in 1928, ended up on display at the North Carolina Museum of Natural Sciences. Note the towing cables wrapped around the middle of the carcass. Photo courtesy of the New Hanover County Public Library, North Carolina Room.

of fish to the area and thereby improved the catch. Meanwhile, back at Wrightsville Beach, the whale's jaw had managed to "disappear," facilitated perhaps by the value of its ivory teeth, and HH would have to find a replacement somewhere.

After the carcass had been buried for more than six months at Topsail Island, the bones were removed and trucked to the fairgrounds in Raleigh, where they were boxed and covered with wet sand for ten more months. Ready or not, however, the bones had to be moved before the state fair opened. Further treatment continued at the museum, including a bath in lye and ammonia, followed by six weeks of drying. Only then, after a year and nine months of smelly labor, waiting, and trouble were the bones assembled with bolts and rods in a process that required another six weeks. Meanwhile, HH had located and purchased a jaw of the right size from the American Museum of Natural History in New York. It was a fortunate find, indeed, but the jaw contained only two teeth, and HH thus had to construct forty-two replacements from dental plaster.

The completed skeleton, suspended from the ceiling on the building's second floor, went on exhibit in early 1930. Still, decades after HH's death, yet one more trouble arose when the skeleton was moved into a new museum facility: the huge skull was too big to get out of the old building, so it was broken up into smaller pieces, which were then reassembled at the new location. Today, the storied skeleton hangs in the museum's Coastal North Carolina Hall and quite literally served as the "backbone" in the design of an earlier version of the museum's logo.

Meanwhile, Clement Brimley also established a significant reputation as a naturalist. Whereas he assumed greater responsibilities for managing their business as HH became involved with the state museum, CS nonetheless pursued his interest in insects and worked on an informal basis with the state's Division of Entomology. In 1919, however, he joined the division as an employee, eventually becoming an associate entomologist and director of the museum's insect collection. His association with the division spanned forty-five years.

CS subsequently expanded the scope of his activities to include other kinds of invertebrates along with amphibians, reptiles, birds, mammals, and plants, and he published dozens of papers on these subjects in scientific journals. Some of these described new taxa of salamanders, turtles, and many insects, primarily those in the orders Hymenoptera (mainly ants, wasps, and bees) and Diptera (flies). Because of this broad knowledge, CS was known among his peers as a "virtuoso naturalist." He wrote *Insects of North Carolina*, coauthored *Poisonous Snakes of the Eastern United States*, and, with his brother and T. Gilbert Pearson (see below), *Birds of North Carolina*. A meticulous collector of data, CS also maintained records of bird migrations in the Raleigh area for forty-six years.

Neither of the brothers attended college, yet both received well-earned recognition for their scientific accomplishments. CS, for example, received an honorary doctorate from the University of North Carolina, and the American Ornithologists' Union elected HH to full membership. Both were among the founders of the North Carolina Academy of Science, for which CS later served as president. The World's Congress of Ornithology enlisted HH as a member of its advisory committee. In honor of the brothers, the North Carolina Museum of Natural Sciences published a journal, *Brimleyana*, between 1979 and 1998. These and other accolades and accomplishments highlight the entry of North Carolina into a new era grounded in science and devoted to the exploration of nature's workings—a transition well led by the brothers Brimley.

T. (Thomas) Gilbert Pearson (1873–1943) also stands tall in the ranks of the state's naturalists. He initially established his credentials as an ornithologist and later spearheaded what became a national movement for protecting birds at a time when hunting was at best weakly regulated. Birds, including sandpipers and egrets, were mercilessly slaughtered en masse for their plumage or to be served as an entrée in high-end restaurants. Waterfowl, too, were prime targets for market hunters, who often shot hundreds of ducks and geese a day at such locations as Chesapeake Bay and Currituck Sound. Much needed to be done to end such carnage, and Pearson—a "mover and shaker" and persuasive orator—was the man to do it.

Although born in Illinois, Pearson spent his boyhood roaming in the woods of central Florida, where his Quaker family worked a citrus orchard. Intrigued by the area's wealth of birds, he began collecting their eggs and soon learned how to mount their skins, both of which broadened his knowledge of ornithology. To offset his tuition, Pearson donated his specimens to the museum at Guilford College, a Quaker school in Greensboro, North Carolina, and agreed to serve as the curator of the school's museum. Field trips to other states produced additional specimens and, in time, resulted in what was then regarded as the largest egg collection in the South. Like those of H. H. Brimley, some of Pearson's specimens were displayed at the Columbian Exposition in Chicago. Indeed, the two men share a lifelong association that focused on natural history and conservation (e.g., both were AOU members).

Pearson was keenly aware of the havoc plume hunters wreaked to the nesting colonies of egrets in Florida; the same was occurring elsewhere as well, but Florida lay at the epicenter of the destruction. Milliners paid high prices for plumes, especially those known as aigrettes, which, ounce for ounce, were worth more than gold! Even worse, the adult birds were easily shot while nesting; hence, their eggs or broods were left to die in the nest. Predictably, such exploitation quickly doomed the colonies. In

1895, the distraught Pearson penned a circular, *Echoes from Bird Land: An Appeal to Women*, which implored fashion-conscious ladies to boycott hats and clothing decorated with bird feathers, which included the stuffed bodies of songbirds that adorned some apparel of the period.

While at Guilford, Pearson also found time to edit the college magazine, manage the baseball team, and become captain of the football team. Following his graduation in 1897, the young biologist worked for the state geologist while seeking another BS degree, which he earned in 1899 from the University of North Carolina at Chapel Hill. He then returned to Guilford College as a professor of biology, but in 1901 he joined the faculty at what is today the University of North Carolina at Greensboro. Pearson kicked off his academic career with a book, *Stories of Bird Life*, which caught the attention of William Dutcher, then chairman of the AOU's committee for the protection of birds. Dutcher proposed that Pearson establish the Audubon Society of North Carolina (ASNC), which would then lobby the state legislature to enact laws that protected birds. Dutcher's committee had previously drafted a "model law" that provided the framework on which to base state legislation. Pearson responded and organized the ASNC in 1902 and became its secretary; his teaching career would soon end.

To his credit, Pearson realized that laws protecting wildlife, no matter how well conceived, must be enforced if they were to accomplish their goals. To that end, he supplemented the AOU framework with language that required enforcement not only for the regulations in the bill but also for the sundry game laws already enacted at the local level in North Carolina. The comprehensive legislation, enacted in 1903, indeed authorized the ASNC to enforce the regulations with its own corps of wardens.[9] It would be known as the Audubon Law and provide a template for

---

9   Audubon wardens faced considerable obstacles, not the least of which was bodily harm at the hands of poachers. Guy Bradley, a warden and deputy sheriff stationed in Florida, was murdered in 1905 when he confronted plume hunters loading their boat with dead birds shot at an egret colony in the Everglades. His death, although tragic, nonetheless provided a catalyst for additional legal protection for Florida's birds. Beginning in 1988, the National Fish and Wildlife Foundation has presented the Guy Bradley Award to a state and federal warden in recognition of their respective achievements.

other states to copy. Remarkably, North Carolina's Audubon Society had emerged as the first state game commission in the South, and Pearson, just twenty-nine, became its first commissioner. Unfortunately, six years later the legislature excluded more than fifty counties from the commission's jurisdiction, thereby thwarting its powers of law enforcement. Meanwhile, Pearson's role had expanded; in 1905 he was elected secretary of the newly formed National Association of Audubon Societies (NAAS). He moved to the national headquarters in New York City in 1911 after Dutcher, founder and first president of the organization, suffered a stroke. When Dutcher died in 1920, Pearson assumed the presidency and remained in that position until 1934.

Pearson tirelessly worked to gain legal protection for birds both nationally and internationally. His efforts contributed to numerous accomplishments, among them a provision in the Tariff Act of 1913 that prohibited the import of plumage from wild birds for commercial uses, the Migratory Bird Treaty (1916) and its enabling act (1918), and the Migratory Bird Conservation Act (1929), which authorized appropriations to establish federal refuges for birds. Under his leadership, the NAAS became the largest organization in the world dedicated to protecting wildlife, including efforts to stem the loss of habitat and the recovery of depleted populations of mammals such as pronghorn antelope. In 1922, Pearson founded the International Committee for Bird Preservation and served as its president until 1938. His visits to Europe and Latin America sprouted new conservation groups and renewed the vigor of those already in existence.

While engaged with the broad sweep of these and similar matters, Pearson also kept his adopted state of North Carolina in mind. With the Brimley brothers as coauthors, he published *Birds of North Carolina* in 1919, which later appeared in two newer editions. The appearance of the book was not without difficulty. The supply of pages, printed and ready for binding, succumbed to a fire at the printers, thereby delaying publication for two years—surely a gut-wrenching event for its authors. He also wrote other books based in large part on his studies of birds in North Carolina, among them *The Bird Study Book*, and edited or contributed to several other volumes about birds. For years he continually pressed the state legislature to replace the ASNC with a state agency with its own wardens, an effort finally realized in 1927 with the creation of the Division of Game and Inland Fisheries, predecessor to the North Carolina Wildlife Resources Commission. In 1924, the University of North Carolina at Chapel Hill awarded Pearson an honorary doctorate in recognition of his unwavering dedication to wildlife conservation. His distinctions include the Medal of the John Burroughs Association, as well as awards from France and Luxembourg.

Pearson contributed to natural history in many ways, especially in the field of ornithology, but he also promoted another dimension: protection of wildlife with laws that were bolstered with rigorous enforcement. His influence extended from the statehouse to the halls of Congress and around the globe—not bad for a young man who traded an egg collection for tuition at a small North Carolina college.

The brilliant red flowers of fire pink set the course for the life work of Bertram Whittier Wells (1884–1978). He identified the striking wildflowers while still in high school, but by his own admission, the moment was one that he would never forget. Wells thus majored in botany at Ohio State University, where he earned both AB and MA degrees in 1911 and 1916, respectively, soon followed by a PhD in ecology—an emerging field at the time—from the University of Chicago. After short tours of teaching at other universities, he accepted the chairmanship of the Botany Department at North Carolina State College (now University) in 1919, a position he held until 1949, although he continued to teach ecology until retiring in 1954.

Early in his career, Wells earned a solid reputation for his work on galls—hardened tumor-like swellings on the stems of herbaceous plants caused by insects or other pathogenic agents—but after arriving in North Carolina he soon redirected his attention to ecology. A statewide compilation of plant communities became his first priority, which he published in 1924, and set the stage for more intensive regional studies to follow. These went beyond a simple listing of the flora and addressed the "how and why" each of these communities developed their distinctive form and structure in their respective environments. One such study tackled the grassy balds that appear as a patchwork on forested mountaintops in western North Carolina; another investigated old-field vegetation in the Piedmont. Others considered maritime forests, the mysterious Carolina bays, and the isolated sandhill region farther inland (all described in the following chapters). Earlier, however, while riding a train to Wilmington in the spring of 1920, Wells saw a community near Burgaw that would remain a centerpiece for his research and commitment to conservation: the Big Savannah, a fifteen-hundred-acre flatland of colorful wildflowers (Fig. 1-8).

Wells, a founding father in America's first generation of ecologists, left a hefty footprint in the course of natural history in North Carolina. His research became the bedrock for future investigations of plant communities throughout the state and remains so today. Still, by most accounts, his *Natural Gardens of North Carolina* stands at the apex of his contributions. The book, first published in 1932 and reprinted in 1967 and again in 2000, presents science in a popular and appealing fashion; uniquely for books of its kind, it introduced wildflowers and other plants in their ecological settings. Wells employed catchy names for several of the chapters, among them "Christmas Tree Land" for the spruce-fir forests of the western mountaintops. As an added feature, the book includes identification keys, supplemented by descriptions, to the prominent species of wildflowers in each community. Wells, ever the conservationist, also took shots at needless exploitation, as when he lamented that "this noble forest [of longleaf pine] is gone—rooted out by hogs, mutilated to death by turpentining, cut down in lumbering, and burned up through negligence" without any "Save the Pines League" working to rescue it. *Natural Gardens* traveled a difficult path toward publication, including delays caused by the Great Depression, but thanks to two stalwart ladies in the Garden Club of North Carolina, the project eventually reached fruition—and natural history in North Carolina remains all the better for it.

In 1938, a then-unknown thirty-one-year-old woman visited Beaufort, North Carolina, where she quickly fell under the spell of Shackleford Banks. The marine life along its shores became the catalyst for a trilogy of books about the sea that began with *Under the Sea-Wind* (1941). Rachel Carson (1907–1964) thus responded to a calling that ultimately

FIGURE 1-8. Ecologist B. W. Wells marveled at the suite of wildflowers that blossomed at the Big Savannah in Pender County, now farmland. The flora at the site once included (*clockwise from upper left*) colic root and orange milkwort, white-tipped sedge, rosebud orchid, and savanna iris.

The Big Savannah: A Treasure of Wildflowers

The Big Savannah that B. W. Wells first saw through a train window quickly became an outdoor laboratory for research, as well as a site for field trips for students enrolled in his ecology class. He investigated the fifteen-hundred-acre savanna in 1925 and 1926 after first completing a state-wide survey of North Carolina's major plant communities in 1924. More than creating a species-by-species listing of the flora, Wells also studied the site conditions that accounted for this unusual tract of vegetation. Chief among these were a regime of fires that recurred every one to six years and a rare soil type that promotes a high water table lasting for ten or more months each year. The savanna, located near Burgaw in Pender County, developed on a band of Liddell soil, a silty loam with a limited distribution in the Coastal Plain; the soil does not occur in any of the surrounding counties. Its fine texture, combined with the flat terrain, effectively curtails drainage. The soggy soil also slows the decomposition of organic matter, but unlike many boggy areas elsewhere, no peat formed at the Big Savannah. In the course of his studies, Wells introduced the term "hydroperiod" to reflect the length of time wetland soils remain saturated; the term continues today as a standard parameter for describing the character of various types of wetland communities.

In *Natural Gardens*, Wells presented with effusive prose the spring-to-fall sequence in which the kaleidoscope of wildflowers appeared at the Big Savannah (Fig. 1-8). He describes, for example, the appearance in mid-April of a white-flowered fleabane whose vast numbers resemble a white tablecloth set with golden goblets—the lingering yellow blooms of trumpet-shaped pitcher plants. By May, he continues, the tablecloth has faded, replaced with the "aristocrats of the plant society," orchids, of which the prominent grass-pink "would attract attention in a Fifth Avenue flower shop." The lavender flowers of pogonias also color the savanna at this season, each "holding out laterally its remarkable crested lip, the landing place of the insect aviator who carries the baggage of pollen." Wells continues the floral march through summer into fall, when, in late November, the "pageant is over" except for an "anti-climax [of] asters, the last offering of the savannah gods. January alone has not a flower to brighten its passing."

Wells also notes the striking absence of legumes and, especially, any species of *Carex*, the largest genus of sedges. Still, the abundance of so many grasses and other taxa of sedges dictates that the Big Savannah be formally designated as a "grass-sedge" bog rather than to identify the site for its array of wildflowers. Among the grasses, toothache grass is especially prominent; florets on the head of this species are distinctively aligned in a single row, and these, when crushed, emit an orange scent that prompted Wells to call it "orange-grass." As Native Americans may have learned, the grass, when chewed, can numb painful gums, hence the common name, toothache grass.

At wetter sites, Wells noted an association between the presence of Le Conte's thistles and ant hills that remained above water during rainy periods. The showy heads of the thistles were sure markers of anthills otherwise hidden by the herbaceous ground cover. Moreover, the "strictly confined" thistles did not share these sites with other plants. The hills were likely those of southern fire ants, a native and less aggressive relative of the highly venomous red imported fire ants that have spread across the United States from Virginia to Texas.

Wells hoped that the Big Savannah might become a state park or receive some other form of protection, but such wishes were never realized. Still, he and others thought that the site would avoid the tractor and plow simply because of its water-logged soils. Thus an offer to sell the property—for two dollars per acre!—brought no response from conservation groups. However, disaster struck in the 1950s when an enterprising farmer from Ohio ditched the bog with a dragline and forever changed its hydrology. Cultivation followed, and Wells's beloved Big Savannah was no more. Or so it seemed.

In 1997, botanist Richard LeBlond discovered a 117-acre tract that mimics the historic Big Savannah, once just five miles distant. It consists of a bog in the rights-of-way under two power lines and an adjoining fire-suppressed woodland of scattered pond pines; the soil is of the same type as that where the Big Savannah developed. Instead of fire, regular mowings of the rights-of-way had retained the characteristic bog vegetation, and neither the woodland nor the utility corridor had ever been farmed. The site hosts at least 170 species of the same native plants that once grew on the Big Savannah. Because of these similarities, LeBlond aptly referred to his discovery as the "Ghost of the Big Savannah." Moreover, the land was for sale, and this time action followed when the North Carolina Land Trust acquired the tract in 2002. Under its stewardship, the trust initiated a three-year cycle of prescribed burns to restore the pine woodland to its original cover of savanna vegetation. Fittingly, the site is formally known as the B. W. Wells Savannah and shall indeed remain a natural garden.

produced *Silent Spring*, a landmark that sparked a new era of environmental awareness that continues today.

Carson's visit to North Carolina was a working vacation. Beaufort was home for a research station operated by the U.S. Bureau of Fisheries, where she became acquainted with ongoing projects, which she would write about after returning to her desk in the bureau's Baltimore office (she was moved to the main office in Washington, D.C.). Several of these projects concerned aquaculture, including one that hoped but failed to establish farms for raising diamondback terrapins, which were declining in the wild from excessive hunting (see Chapter 3). In addition to Shackleford Banks, Carson roamed about on Town Marsh, a small island that decades later would be part of a preserve named in her memory. She carried a small notebook, in which she entered a tangle of notes about what she saw and felt on the North Carolina coast. The result were lines often enriched with color, among them "the crests of the waves, just before they toppled, caught the gold of the setting sun then dissolved in a mist of silver"—eloquence stands as a hallmark of her writings.

A decade earlier, Carson had attended the Pennsylvania College for Women (now Chatham College), where, after initially selecting English for a major, she graduated with a degree in biology in 1929. The change in part was prompted by a concluding line in a Tennyson poem, "For the mighty wind arises, roaring seaward, and I go." Prior to her senior year, she earned a scholarship that allowed her to take a summer course at the renowned Marine Biology Laboratory in Woods Hole, Massachusetts, and for the first time saw the ocean. In 1932, she earned an MS degree in biology from Johns Hopkins University with the intent of continuing for a PhD. In the midst of the Great Depression, however, the lack of funds ended her formal education (years later she received two honorary doctorates), and she began working for the U.S. Bureau of Fisheries, first in a temporary position but in 1936 as a full-time employee with the title of junior aquatic biologist, then a rare opportunity for a woman.[10] Her

10 In 1940, the Bureau of Fisheries, a unit in the Commerce Department, and the Bureau of Biological Survey in Agriculture were merged to become the U.S. Fish and Wildlife Service in the Department of the Interior. Carson maintained her position during the merger.

responsibilities focused on turning field data into reports, writing brochures on conservation issues, and, on occasion, doing some lab work that involved determining the age of fishes.

Despite favorable reviews—"a beautiful and unusual book," according to one critic—*Under the Sea-Wind* was not a commercial success, largely because World War II broke out just as the book was released.[11] In her words, "the rush to the book store that is the author's dream never materialized." Still, her remarkable writing skills were not lost on editors at leading publishers and eventually led to two more books that also focused on marine life: *The Sea Around Us* (1951) and *The Edge of the Sea* (1955); both were financial successes. Indeed, the former remained on the *New York Times*' best-seller list for eighty-six weeks, a success that enabled Carson to leave her government job to become a full-time writer. But an even greater triumph, both financially and in its impact, was still to come.

In 1962, Carson published *Silent Spring*, a blockbuster that sold sixty-five thousand copies in the first two weeks. The book examined at length the misuse of pesticides and its devastating effects on wildlife. Its genesis began after World War II, when DDT showed up in milk originally collected to detect the presence of radioactive fallout from widespread nuclear testing. Soon other evidence emerged, not the least of which were declining populations of birds, especially those of brown pelicans, ospreys, and others at the peak of food chains, known to ecologists as apex predators. In these situations, DDT, DDE, and its sister compounds accumulated as they passed upward in the chain. This process, known as biomagnification, begins with relatively harmless applications (often just a pound or so per acre), but their impact steadily grows stronger each time an organism continues to consume contaminated food (i.e., big fish eating many small fish, followed by a pelican consuming many big fish). The effect was insidious. Whereas dead and dying birds were rarely evident, the contaminated birds laid thin-shelled eggs that cracked during incubation, with the obvious result that their

---

11   When reissued in 1952, *Under the Sea-Wind* became a best seller and remains in print.

populations declined. In some cases, however, pesticides indeed killed numbers of birds, events that became Carson's touchstone—a spring without singing birds.

Predictably, *Silent Spring* ignited an outcry, sometimes quite vicious, from the chemical and agricultural industries, as well as from the politicians that represented those interests. But it also galvanized conservationists, whose voices became loud and clear—harmful pesticides and their use must be regulated. Overnight, Carson had become a champion for a healthy environment. John F. Kennedy, for example, asked her to appear before the President's Science Advisory Committee, and in no small way, her book led to the formation of the Environmental Protection Agency.

Rachel Carson died of cancer in 1964, some say caused by the very chemicals she highlighted in *Silent Spring*. She had initiated a social revolution that garnered wide recognition, including a National Wildlife Refuge in Maine and a marine preserve in North Carolina named in her honor. Each year the National Audubon Society presents its Rachel Carson Award to an American woman whose work has significantly advanced conservation. Bridges and schools bear her name, as do science and literature prizes, as well as two seagoing research vessels. She received, posthumously, the Presidential Medal of Freedom in 1980. Her home in Maryland, where she wrote *Silent Spring*, is a National Historic Landmark, and her image has appeared on a postage stamp. Perhaps most of all, in 1972 DDT was among the pesticides banned from further use in the United States. Rachel Carson once left her footprints on the beaches of North Carolina, but she cast a far more permanent impression on the quest for a safe and clean environment.

It seems fitting to close with another observation by John Lawson, who, despite his own pioneering contributions to the natural history of North Carolina, well knew that he had described only a small part of the region's biological wealth. In his words, "There will be a great number of Discoveries made by those that shall come hereafter [and] when another Age is come, the Ingenious then in being may stand upon the Shoulders of those that went before them, adding their own Experiments to what was delivered down to them by their Predecessors, and then there will be something towards a complete Natural History." Others, as we have seen, indeed did follow—and continue to do so.

READINGS AND REFERENCES

General Sources

Blevins, D., and M. P. Schafale. 2011. Wild North Carolina: Discovering the Wonders of Our State's Natural Communities. University of North Carolina Press, Chapel Hill. [Features elegant photos with brief but engaging commentaries about prominent environments across the state.]

Earnhardt, T. 2013. Crossroads of the Natural World: Exploring North Carolina with Tom Earnhardt. University of North Carolina Press, Chapel Hill. [A tour de force of the state's natural wonders.]

Fussell, J. O., III. 1994. A Birder's Guide to Coastal North Carolina. University of North Carolina Press, Chapel Hill. [Includes brief site descriptions of much of the state's coastline.]

Simpson, B. 1997. Into the Sound Country: A Carolinian's Coastal Plain. University of North Carolina Press, Chapel Hill. [A regional portrait of eastern North Carolina.]

The Landscape

Allen, R. P., and R. T. Peterson. 1936. The hawk migrations at Cape May Point, New Jersey. Auk 53:393–404. [This seminal study of migratory behavior provides a large-scale example of a similar event at Fort Fisher.]

Byrd, B. L., A. A. Hohn, G. N. Lovewell, et al. 2014. Strandings as indicators of marine mammal biodiversity and human interactions off the coast of North Carolina. Fishery Bulletin 112:1–23.

Davis, R. J., and J. F. Parnell. 1983. Fall migration of land birds at Fort Fisher, New Hanover County, North Carolina. Chat 47:85–95.

Kahl, M. P. 1964. Food ecology of the wood stork (*Mycteria americana*) in Florida. Ecological Monographs 34:97–117.

LeBlond, R. J. 2001. Endemic plants of the Cape Fear Arch region. Castanea 66:83–87.

Leonard, S. W., and R. J. Davis. 1981. Natural areas inventory for Pender County, North Carolina. Coastal Energy Impact Report No. 11, North Carolina Department of Natural Resources and Community Development, Raleigh. [See pages 145–146 for more about the wet marl forest community.]

Parnell, J. F., and T. L. Quay. 1962. The populations, breeding biology, and environmental relations of the black duck, gadwall, and blue-winged teal at Pea and Bodie Islands, North Carolina. Proceedings of the Annual Conference, Southeastern Association of Fish and Wildlife Agencies 16:53–67.

Schafale, M. P. 2020. Sandy pine savanna (rush featherling subtype). *In* Classification of the Natural Communities of North Carolina, 4th Approximation. In preparation. North Carolina Natural Heritage Program, Department of Natural and Cultural Resources, Raleigh. Accessed March 5, 2020.

Schweitzer, S. 2019. A rare bird comes to North Carolina. Wildlife in North Carolina 83(3):16–21.

Searles, R. B. 1984. Seaweed biogeography of the mid–Atlantic coast of the United States. Helgoland Marine Research 38:259–271.

Sorrie, B. A., and A. S. Weakley. 2001. Coastal Plain vascular plant endemics: phytogeographic patterns. Castanea 66:50–82.

Thornhill, R., A. Krings, D. Lindbo, and J. Stucky. 2014. Guide to the vascular flora of the savannas and flatwoods of Shaken Creek Preserve and vicinity (Pender & Onslow Counties, North Carolina, USA). Biodiversity Data Journal 2:e1099.

Webster, W. D., W. W. Golder, A. Neville, and J. F. Gouveia. 1987. The distributional status of the eastern wood rat, *Neotoma floridana floridana*, in North Carolina. Journal of the Elisha Mitchell Scientific Society 103:111–112.

Wells, B. W. 1928. Plant communities of the Coastal Plain of North Carolina and their successional relations. Ecology 9:230–242.

Wiedner, D. S., P. Kerlinger, D. A. Sibley, et al. 1992. Visible morning flights of neotropical landbird migrants at Cape May, New Jersey. Auk 109:500–510.

A Wonderland for Naturalists

Berkeley, E., and D. S. Berkeley. 1982. The Life and Times of John Bartram: From Lake Ontario to the River Saint Johns. University Presses of Florida, Tallahassee.

Bolen, E. G. 1994a. The Bartrams in North Carolina. Wildlife in North Carolina 60(5):16–21.

Bolen, E. G. 1994b. Bird man of Fort Macon. Wildlife in North Carolina 58(11):16–19.

Bolen, E. G. 1998. John Lawson's legendary journey. Wildlife in North Carolina 62(12):22–27.

Bolen, E. G. 2006. Mr. Wilson's woodpecker. Wildlife in North Carolina 70(5):26–31.

Brickell, J. 1968. The Natural History of North Carolina. Johnson Publishing, Murfreesboro, N.C. [Reprint of the original 1737 work published in Dublin, Ireland, of which much was "borrowed" directly from John Lawson's 1709 book, *A New Voyage to Carolina.*]

Broadhead, M. J. 1971. Elliott Coues and the sparrow war. New England Quarterly 44:420–432.

Catesby, M. 1747. Of birds of passage. Philosophical Transactions, Royal Society of London 44:435–444.

Catesby, M. 1974. The Natural History of Carolina, Florida, and the Bahama Islands. Beehive Press, Savannah, Ga. [Reprint of the third edition of Catesby's two-volume work published, respectively, in 1732 and 1743.]

Cutright, P. R., and M. J. Broadhead. 1981. Elliott Coues, Naturalist and Frontier Historian. University of Illinois Press, Urbana.

Feduccia, A., ed. 1985. Catesby's Birds of Colonial America. University of North Carolina Press, Chapel Hill. [Updates Catesby's text with current knowledge for each bird described in the original book.]

Fitzpatrick, J. W., M. Lammertink, M. D. Lundeau Jr., et al. 2005. Ivory-billed woodpecker (*Campephilus principalis*) persists in continental North America. Science 308:1460–1462. [A sighting received by many with considerable skepticism.]

Frick, G. F., and R. P. Stearns. 1961. Mark Catesby: The Colonial Audubon. University of Illinois Press, Urbana.

Giroux, J.-F., G. Gauthier, G. Costanzo, and A. Reed. 1998. Impact of geese on natural habitats. Pages 32–57 *in* The Greater Snow Goose: Report of the Arctic Goose Habitat Working Group (B. D. J. Batt, ed.). Arctic Goose Joint Venture Special Publication. U.S. Fish and Wildlife Service, Washington, D.C., and Canadian Wildlife Service, Ottawa.

Hoffman, N. E., and J. C. Van Horne, eds. 2004. America's Curious Botanist: A Tercentennial Reappraisal of John Bartram, 1699–1777. American Philosophical Society, Philadelphia, Pa.

Holloman, C. R. 1991. Lawson, John. Pages 34–36 *in* Dictionary of North Carolina Biography, Vol. 4 (W. S. Powell, ed.). University of North Carolina Press, Chapel Hill. [Holloman deserves credit for discovering documents that filled a long-standing void in Catesby's biography.]

Huler, S. 2019. A Delicious Country: Rediscovering the Carolinas along the Route of John Lawson's 1700 Expedition. University of North Carolina Press, Chapel Hill.

Hulton, P. 1985. America 1585: The Complete Drawings of John White. University of North Carolina Press, Chapel Hill.

Hume, E. E. 1978. Ornithologists of the United States Army Medical Corps. Arno Press, New York.

Jackson, J. A. 2006. Ivory-billed woodpecker (*Campephilus principalis*): hope, and the interface of science, conservation, and politics. Auk 123:1–15. [Responds to the claim that the birds persist in continental North America.]

Lawson, J. 1967. A New Voyage to Carolina (H. T. Lefler, ed.). University of North Carolina Press, Chapel Hill. [A facsimile of the 1709 edition of Lawson's book, with an informative introduction by the editor; but see Holloman (1991) for biographical material unknown in 1967.]

Magee, J. 2010. The Art and Science of William Bartram. Pennsylvania State University Press, University Park.

Nelson, E. C., and D. J. Elliot, eds. 2015. The Curious Mister Catesby: A "Truly Ingenious" Naturalist Explores the New Worlds. University of Georgia Press, Athens.

Philippe, M. 2011. How many species of *Araucaryoxylon*? Comptes rendus palevol 10:201–208. [Reviews difficulties associated with the taxonomy of petrified wood.]

Quinn, D. B. 1985. Set Fair for Roanoke: Voyages and Colonies, 1584–1606. University of North Carolina Press, Chapel Hill.

Simpson, M. B., Jr., and D. S. McAllister. 1986. Alexander Wilson's southern tour of 1809: the North Carolina transit and subscribers to the *American Ornithology*. North Carolina Historical Review 63:421–476. [Includes Wilson's experience in Wilmington with an ivory-billed woodpecker.]

Smith, H. M., M. J. Preston, R. B. Smith, and E. F. Irey. 1990. John White and the earliest (1585–1587) illustrations of North American reptiles. Brimleyana 16:119–131.

Smith, T. J., III, and W. W. Odum. 1981. The effects of grazing by snow geese on coastal salt marshes. Ecology 62:98–106.

Stewart, D. 1997. Mark Catesby. Smithsonian 28(6):97–103.

Tucker, A. 2008. Sketching the earliest views of the New World. Smithsonian 39(9):48–53.

Yih, D. 2013. Mark Catesby: pioneering naturalist, artist, and horticulturist. Arnoldia 70(3):25–36.

Into the Twentieth Century

Carson, R. 1942. Under the Sea-Wind: A Naturalist's Picture of Ocean Life. Simon and Schuster, New York.

Carson, R. 1962. Silent Spring. Houghton Mifflin, Boston.

Cooper, J. E. 1979. The brothers Brimley: North Carolina naturalists. Brimleyana 1:1–14.

Cotrufo, M. 2007. "Trouble." http:/files.naturalsciences.org/Trouble.pdf. [A definitive account of the whale's journey from beach to museum, with numerous photos.]

Earley, L. S. 1989. The natural gardens of B. W. Wells. Wildlife in North Carolina 53(2):18–23.

Enderson, J. H., and D. D. Berger. 1970. Pesticides, eggshell thinning, and lowered production of young in prairie falcons. BioScience 20:355–356.

Martin, M. 2001. A Long Look at Nature: The North Carolina State Museum of Natural Sciences. University of North Carolina Press, Chapel Hill.

McIver, S. B. 2003. Death in the Everglades: The Murder of Guy Bradley, America's First Martyr to Environmentalism. University of Florida Press, Gainesville.

Odum, E. P. 1949. A North Carolina Naturalist, H. H. Brimley: Selections from His Writings. University of North Carolina Press, Chapel Hill.

Orr, O. H., Jr. 1992. Saving American Birds: T. Gilbert Pearson and the Founding of the Audubon Movement. University of Florida Press, Gainesville.

Orr, O. H., Jr. 1994. Thomas Gilbert Pearson. Pages 52–54 in Dictionary of North Carolina Biography, Vol. 5 (W. S. Powell, ed.). University of North Carolina Press, Chapel Hill.

Pearson, T. G. 1917. The Bird Study Book. Doubleday, Garden City, N.Y.

Pearson, T. G. 1937. Adventures in Bird Protection: An Autobiography. D. Appleton-Century, New York.

Pearson, T. G., C. S. Brimley, and H. H. Brimley. 1919. Birds of North Carolina. Edwards and Broughton, Raleigh, N.C. [The first of three editions.]

Potter, E. F. 1986. H. H. and C. S. Brimley: brother naturalists. Chat 50:1–9.

Rudd, R. L. 1964. Pesticides and the Living Landscape. University of Wisconsin Press, Madison.

Simpson, M. B., Jr. 1979a. Brimley, Clement Samuel. Pages 227–228 *in* Dictionary of North Carolina Biography, Vol. 1 (W. S. Powell, ed.). University of North Carolina Press, Chapel Hill.

Simpson, M. B., Jr. 1979b. Brimley, Herbert Hutchinson. Page 228 *in* Dictionary of North Carolina Biography, Vol. 1 (W. S. Powell, ed.). University of North Carolina Press, Chapel Hill.

Souder, W. 2012. On a Farther Shore: The Life and Legacy of Rachel Carson. Crown, New York. [A major source; see pages 86–95 for details about her visit to North Carolina and first book.]

Spitzer, P. R., R. W. Risebrough, W. Walker, et al. 1978. Productivity of ospreys in Connecticut–Long Island increases as DDE residues decline. Science 202:33–335.

Troyer, J. R. 1986. Bertram Whittier Wells (1884–1978): a study in the history of North American plant ecology. American Journal of Botany 73:1058–1078.

Troyer, J. R. 1993. Nature's Champion: B. W. Wells, Tar Heel Ecologist. University of North Carolina Press, Chapel Hill. [A primary source for material discussed here.]

Wells, B. W. 1967. The Natural Gardens of North Carolina. University of North Carolina Press, Chapel Hill. [A reprint of the original 1932 volume.]

Wolfe, D. A. 2000. A history of the Federal Biological Laboratory at Beaufort, North Carolina, 1899–1999. U.S. National Oceanic and Atmospheric Administration, Washington, D.C.

Infobox 1-1: A New Taxonomy

Blunt, W. 2001. Linnaeus: The Compleat Naturalist. Princeton University Press, Princeton, N.J.

Infobox 1-2: The Big Savannah: A Treasure of Wildflowers

Shelingoski, S., R. J. LeBlond, J. M. Stucky, et al. 2005. Flora and soils of Wells Savannah: an example of a unique savanna type. Castanea 70:101–114. [Note: the two spellings in this and other titles cited here reflect a change from the early twentieth-century "savannah" used by Wells—and by Colonial-era Georgians—to the preference of today's ecologists for "savanna."]

Wall, W. A., T. R. Wentworth, S. Shelingoski, et al. 2011. Lost and found: remnants of the Big Savannah and their relationship to wet savannas in North Carolina. Castanea 76:348–363.

Wells, B. W., and I. V. Shunk. 1928. A southern upland grass-sedge bog: an ecological study. North Carolina Agricultural Experiment Station Technical Bulletin No. 32. Raleigh.

II

Beaches, Dunes, and Barrier Islands

North Carolina's Outer Banks stand as the best-known barrier islands in North America, competing only with Padre Island, the pride of Texas. Both are summer playgrounds not only for throngs of sun-loving tourists but also for those with binoculars, field guides, and a bent for natural history. Nesting sea turtles, hardy vegetation, busy crabs, and a wealth of birds provide naturalists with a menu of attractions. In season, a lucky observer might even see a northern right whale, a species now struggling for survival. For all, the edge of the sea is indeed a dynamic environment of restless sand and eternal surf.

Anatomy of the Outer Banks

Formations known as barrier islands parallel much of the Atlantic coastline from Massachusetts to Texas, with those in North Carolina among the more prominent for their dynamic geology, thanks in large measure to the prevalence of nor'easters and hurricanes (see below). Barrier islands at times morph into peninsulas when storm-driven sands fill in the intervening gap with the mainland. Storms likewise cut inlets (or passes) through peninsulas to create (or re-create) new barrier islands or to bisect a larger island into smaller units. Many of these changes

FIGURE 2-1. Storms produce overwashes that carry sand from the seaside edge of a barrier island to its inner edge, which gradually causes the island to migrate landward during times of rising sea levels. In 2011, Hurricane Irene produced the washover shown here on South Core Banks.

persist for decades or even centuries, whereas others last only until the next storm. Some of the longer-lasting channels, named when they were formed, still appear decades later on maps and charts as "New Inlet." Most barrier islands formed three thousand to seven thousand years ago, but the process leading to their formation remains uncertain and, indeed, may involve more than one mechanism. Most, however, associate their beginnings with the advent of rising sea levels.

Rising sea levels continue to exert an important influence on barrier islands—they migrate landward in what might be described as storm-driven rollover. In short, overwashes transport sand from the beachfront to the inner edge of the island so that an island steadily rolls over itself from seaside to soundside (Fig. 2-1). As an island migrates landward, the sand buries marsh vegetation, which later emerges as peat when the former inner edge of the island eventually becomes the beachfront. In some cases, the stumps of trees that once grew along the inner edge of an island also appear on the beachfront; such a "stump field" can be seen on the beachfront at Currituck Banks. Barrier islands thus migrate landward when sea levels rise but remain inland as sandy ridges when sea levels fall. In North America, this phenomenon occurs primarily on the Atlantic coast because of the gentle contour of the continental shelf, whereas the shelf along the Pacific coast drops abruptly into deep water.

THE SWASH ZONE

As waves greet a sandy shore, they "break" when the onrushing mass of water can no longer support itself in the shallows, a turbulence that marks the surf zone. A thin sheet of water continues moving up the beach, steadily losing its energy; then, with its momentum spent, the flow reverses direction and retreats to the surf zone. The perpetual upwash and backwash of dying waves define the swash zone, where waves die with the rhythm of a pendulum.

At first glance, the swash zone seems devoid of life, but in fact it supports a species-rich community known as meiofauna (commonly, diatoms, nematodes, crustaceans, and other minute invertebrates), which thrives in the spaces between sand grains. Larger organisms also live in the swash zone, where they quickly burrow when exposed by the moving

FIGURE 2-2. Swash zones provide foraging areas for shorebirds and gulls. Sanderlings, the small birds shown here, rush to keep pace with the advance and retreat of the waves.

water and hence go unnoticed. These include filter feeders such as surf clams, polychaetes, and sand fleas. Numerous shorebirds, particularly sandpipers, thrive on the hidden riches of the swash zone (Fig. 2-2). In movements that suggest windup toys, sanderlings and other short-legged species forage on food exposed by the backwash, then hastily reverse course when the next wave arrives.

Coquina clams, about the size of a thumbnail, abound in the swash zones of undisturbed beaches, where their presence serves as a biological indicator of a healthy shore. They partially emerge during the upwash, filter a meal of single-celled algae and detritus carried in the backwash, then disappear again into the sand in a cycle that renews itself with each new wave. Coquinas also move up and down the beach profile (slope) as the tide changes and thereby stay within the boundaries of the swash zone. In good habitat, coquina populations may reach densities of several hundred or more per square yard. Indeed, their abundance during the Pleistocene produced a soft limestone rock that incorporated clusters of their shells (see below). With a little effort, gourmands can acquire enough coquinas as stock for a savory broth or chowder.

Crustaceans that are the size of a thimble and known as sand fleas or mole crabs—but that are neither flea nor crab—live at the lower edge of the swash zone (Fig. 2-3). Like coquinas, they move up and down the slope in step with the changing tides; they likewise partially emerge with an incoming wave, feed as the wave retreats, then instantly return to safety under the sand. Sand fleas nonetheless give away their hiding place. To capture food in the backwash, they erect feathery antennae that create a telltale V-shaped wake in the receding water that becomes a signpost for hungry shorebirds, as well as for anglers in search of bait. The effect is similar to the wake formed by an extended periscope that pinpoints the location of a submarine. In a twist from most other invertebrates (e.g., fiddler crabs), sand fleas indiscriminately ingest sand and food together, leaving their internal organs to separate the two.

Sand fleas burrow tail first, digging with their telsons (or "tails"), which also provide an anchor for holding fast in their burrows. Their telsons,

FIGURE 2-3. Small crustaceans known as sand fleas or mole crabs (*left*) thrive hidden in the swash zone, as do colorful coquina clams (*right*); both form links in the shoreline food chain. Photo credit: Pernille Ruhalter (*left*).

when curled, protect the soft underbelly of sand fleas and shelter the egg masses carried on the abdomens of females. Although a bit intimidating in appearance, sand fleas lack claws on any of their five pairs of legs, nor can they bite or sting. Their rear legs function as paddles that enable sand fleas to swim in turbulent water, but on firmer footing they scuttle backward. In winter, sand fleas leave the beach and head for the bottom in deeper waters.

THE WRACK LINE

A linear accumulation of organic materials—the wrack line—tossed by storm-driven waves may form a few feet beyond the upper limits of the normal high-tide line (Fig. 2-4). These materials, primarily seagrasses and the occasional remains of a hapless sea creature, provide habitat for numerous invertebrates and the foraging grounds for gulls and other scavengers. Shorebirds likewise seek food in wrack lines. Additionally, as the debris decays, it nourishes the beach with much-needed nitrogen. Wrack lines also attract the curious attention of beachcombers. Many delight in the weathered charm of driftwood, which may have spent years at sea before washing ashore and include the remains of a doomed vessel—perhaps, with some imagination, from a dhow wrecked on an African shore. Whatever its origins, driftwood often provides habitat for some interesting creatures, not the least of which are shipworms. Hardly worms, these instead are uniquely adapted mollusks with two shells, each like the unhinged half of a small clam, positioned at the anterior end of their tubular bodies. The shells provide shipworms with boring tools that throughout history have damaged wooden ships and piers. Driftwood in the wrack line, although long barren of living shipworms, may be riddled with their telltale tunnels. In life, they may be a foot long, and, with the aid of cellulose-digesting bacteria, they feed primarily on the wood they excavate. Driftwood also attracts barnacles, sometimes including gooseneck barnacles, which gave rise to a medieval myth: their long peduncles and feathery feeding structures seemed to imaginative English naturalists like preformed geese ready to hatch. Ergo, the birds disappeared each spring because they turned into barnacles. The geese, of course, had migrated to their northern nesting areas, but the fanciful explanation still persists in birds known as barnacle geese.

FIGURE 2-4. The accumulation of organic matter in the wrack line provides food for a variety of scavengers ranging from foxes to ghost crabs. Beachcombers likewise search the wrack line for driftwood and other "treasures."

Armor-Clad Shorelines

Sandy beaches naturally erode when longshore drift moves wave-stirred sand along the beach. In North Carolina, this moves sand north to south on east-facing beaches; hence, barrier islands such as Hatteras Island expand at their southern ends and diminish at their northern ends. On Shackleford Banks and other south-facing beaches the movement is west to east. Barrier spits similarly increase or decrease in length depending on their orientation. Inlets also migrate with the shifting sands. Storms, of course, accelerate this process and commonly leave behind foreshortened beaches with cliff-like berms. In time, erosion threatens—and eventually claims—beachfront structures of all sorts, including lighthouses. Moreover, rising sea levels currently increase the severity and occurrence of beachfront erosion.

In response, structures designed to hold sand in place by blocking longshore drift now "harden" many shorelines. Rows of groins constructed with boulders, concrete, or steel jut outward from the shore into the ocean like the tines of a giant rake; in North Carolina, these groins trap sand on their northern or eastern side, but erosion continues on the other, which results in a saw-tooth shoreline of "have and have not" beachfronts. Additionally, when strong longshore currents generated by storms hit the groins they change direction and carry huge amounts of sand offshore. Hence, the shoreline loses sand, and the sea floor is disturbed. Rock or concrete projections known as jetties protect the openings of inlets, especially those originally cut with dredges. Designed to prevent the channels from filling in, jetties also accumulate sand on one side while starving the beach on the other. Seawalls, built along the shoreline to thwart storm surges, likewise impact adjacent beaches. Storm-driven waves that would normally dissipate their energy as they sweep across the swash zone now thunder against the wall, where they form a massive outwash of turbulent water that carries sand offshore. In time, beaches in front of these structures may entirely disappear, leaving the ocean lapping against the foot of the seawall. In all, about 14 percent of the coast in the United States has been hardened, as has more than 6 percent of the coast in North Carolina.

Whatever pluses these and other structures (e.g., huge sand bags) may have, they also present environmental negatives. When groins or jetties cause a narrowing of a beach, often to half its original width, habitat disappears for the meio- and the macrofauna (respectively, small to somewhat larger invertebrates) in the surf and swash zones. Such disturbances, in turn, rupture the food chain for birds and small fishes foraging in these areas. Indeed, when compared with unmodified beaches, fewer (by half) species and (by two-thirds) numbers of shorebirds forage along hardened shorelines. Wrack accumulations also diminish on hardened beaches, which deprives ghost crabs and other invertebrates of food and cover, as well as impairing nutrient recycling on beaches. Because hardened shorelines are often renourished with sand dredged from other areas, commonly from sites offshore, the transferred material may differ from the original sands in ways (e.g., large or smaller grains) that have ecological consequences, among them difficulties for nesting sea turtles. Although these outcomes remain unproved, the changes wrought by shore-protection structures may create sites where invasive species might find suitable niches and compete with the native fauna.

In 1985, North Carolina's Coastal Resources Commission recommended banning further hardening of the state's beaches, with exceptions for protecting historic structures or navigation channels. The ban, although lacking the force of law, nonetheless halted further hardening of the coast for more than fifteen years. Finally, in 2003, the General Assembly unanimously voted to transform the ban into law. Still, special interests seek ways to circumvent or weaken the legislation, which commands constant vigilance from those wishing to preserve the natural dynamics and integrity of the state's beaches.

As coastal botanist Paul Hosier aptly notes, "Sea oats is a botanical superhero." And indeed it is. Its adaptations to the harsh conditions encountered on the foredunes include the ability to survive sand burial, excessive heat, drought, and exposure to salt aerosols, as well as short-term flooding by storm-driven seawater (Fig. 2-5). Ecologically, sea oats takes over the same niche occupied by American beachgrass in dune systems along coastlines farther north: both species build and stabilize dunes. These hardy grasses meet and overlap in a broad ecotone that extends between Cape Hatteras and southern Virginia, although stragglers of each extend, respectively, even farther north or south of this ecotone; the range of sea oats thereafter continues southward to the Gulf of Mexico and beyond.

Sea oats gives rise to an unmistakable inflorescence—a seed head, or panicle, of flattened inch-long spikelets that turn to the color of honey as they mature. (And while they are an attractive garnish for floral arrangements, neither the plants nor their panicles can be legally collected in North Carolina without the consent of the landowner.) Somewhat oddly, the numerous spikelets produce few seeds, and, for the most part, the plants instead rely on their scaly-looking rhizomes to spread across the dunes.

Sea oats also forms nodes at the base of its stems; these commonly produce roots when sand accumulates around the plant, further anchoring the plant and stabilizing the dune. In fact, as sand begins to bury the plants it stimulates additional aboveground growth, as well as further expansion of their rhizome and nodal root systems. Like its northern counterpart, sea oats can be damaged by human traffic and therefore compromise the integrity of the dune itself; hence, controlled access through the dunes on well-marked paths becomes a basic tool for the protection of coastal shorelines.

Seacoast bluestem likewise helps stabilize dunes and shares all but one of the adaptations of sea oats: it cannot survive sand burial. Accordingly, it favors sheltered locations on dunes lying behind those immediately facing the open beachfront.

FIGURE 2-5. Sea oats serves as the primary dune-building grass on beach fronts along the North Carolina coast and, as shown here, provides a striking backdrop for a colony of black skimmers.

FIGURE 2-6. The endemic crystal skipper feeds on the nectar of coastal morning glory at Bogue Banks. Photo credit: Becky Harrison, U.S. Fish and Wildlife Service.

Seacoast bluestem plays another role as well. An isolated population of what may be a new species of butterfly—the crystal skipper—occurs on Bogue Banks, where it is confined to a dunes system lying between two natural barriers, an inlet and a stand of maritime forest. As adults, the butterflies feed on nectar from various wildflowers, but coastal bluestem serves as the only food source for the larvae (Fig. 2-6). Two generations emerge each summer, with the chrysalis (cocoon) of the second overwintering in a delayed state of development until the following spring. So far as now known, the host-specific relationship between the larvae and the plant does not provide any chemical protection against predators akin to the well-known example between monarch butterflies and milkweed. Fortunately, two protected areas—Bear Island at Hammock State Park and Fort Macon State Park—provide a secure core of habitat for crystal skippers, which are currently designated as a species of special concern.

Dune hairgrass is among the other grasses growing on dunes, as well as on marsh edges and in slacks (see below). Known locally as "sweetgrass," it provides the fiber for weaving coiled baskets, a distinctive craft imported by slaves from West Africa and still practiced by their descendants living on the coast of South Carolina; the plants were not abundant enough for the tradition to develop in North Carolina. Dune hairgrass grows in clumps of blue-green foliage and tall stems topped in season by hairlike plumes of pink to reddish inflorescences. In fall, the plants offer a stunning sight on the Outer Banks.

Final mention goes to marsh hay, also known as saltmeadow cordgrass, which helps anchor the low, broader dunes that form behind the steeper foredunes facing the beachfront. As suggested by its name, marsh hay also forms extensive meadows known in northern states as "high marsh" that develop just above the tidal line.

Only a few hardy plants thrive where the beach meets the dune system, among them seabeach amaranth. Others seek dunes better protected from direct contact with the oceanfront; sea rocket, southern seaside goldenrod, and beach bean represent this group (Fig. 2-7).

FIGURE 2-7. Dune vegetation includes (*clockwise from upper left*) sea rocket, beach bean, seabeach amaranth, and southern seaside goldenrod. Photo credit: Kevin Knutsen (*bottom right*).

Unfortunately, an aggressive plant native to the Pacific Rim (e.g., Korea and Indonesia) has invaded dune systems along the Atlantic and Gulf coasts. Imported in the 1980s as an ornamental and a shoreline stabilizer, beach vitex can smother an entire dune with a dense, shrubby mat that crowds out sea oats and other native vegetation (Fig. 2-8). Clearly salt tolerant, the plants produce large quantities of seeds, as well as runners that extend for distances up to sixty feet; these sprout roots at their nodes that initiate and nourish more plants. The grapelike seeds float, as do storm-broken fragments of the plant itself; both disperse to nearby beaches, where they establish yet another colony. The mats take over the niche occupied by a threatened species—seabeach amaranth—and sites where sea turtles nest. In response, coastal states, including North Carolina, have organized task forces to remove beach vitex wherever possible.

FIGURE 2-8. Beach vitex, an invasive species from Pacific shorelines, crowds out native dune vegetation at many locations along the coastlines of North Carolina and other southeastern states.

Sites between dunes, called slacks or sometimes swales, offer local conditions where vegetation can sometimes further develop (i.e., through stages in a process called succession). Initially, wax myrtle, at times accompanied by yaupon, establishes dense, low-growing patches known as maritime thickets. These seldom include other woody vegetation, although bayberry, a close relative, may replace wax myrtle as the dominant on the northern coastline of North Carolina. Vines such as greenbriar and Virginia creeper often weave into the dense fabric of the canopy, but the limited amount of light penetrating the closed canopy otherwise prevents the growth of herbaceous ground cover. Salt aerosols shape the canopy into a slope extending from ground level at the windward edge to a peak on the leeward side, a process likewise sculpting maritime forests (see below). The edges of these thickets provide sites where salt-tolerant grasses often gain footholds. One of these, maritime bushy bluestem, produces large, silvery, and seed-rich inflorescences not unlike small feather dusters (which to some suggested an alternate name, bushy broomgrass).

In time, the thickets merge, and the underlying sandy soil becomes enriched with leaf litter and other organic matter; the shrubby growth also traps windblown silt and clay particles that hasten the development of fertile soil atop the sand. At this point, coastal red cedar adds to the complexity of the vegetation and heralds the beginnings of a maritime forest; live oak, the signature species of the climax community, soon follows (see below). Not all slacks, of course, develop maritime forests—dune systems are too unstable for such consistency. Storms and blowouts can easily alter the course of succession in a slack and repeatedly set back the sequence to "square one." In some cases, a slack may retain surface water and become a semipermanent wetland of cattails and other freshwater marsh vegetation.

Whales and Whaling in North Carolina

North Carolina's whaling industry differed markedly from its more famous counterparts in New England: hardy whalers operated from shore on a day-to-day basis, not from a fleet of ships on long voyages to far-flung hunting grounds, like those famously sailing from Nantucket and New Bedford. Instead, most North Carolina whaling sites were unnamed locations now erased by time and storm; virtually all were on Bogue and Shackelford Banks. Except for Beaufort, most of these were rude camps, not busy towns with wharfs, ship chandlers, and the prominent homes of sea captains, along with some less savory accommodations. A high point near each camp served as a "crow's nest" where a lookout watched for whales passing offshore. (One site, then named Lookout Woods, was the forerunner of the ill-fated Diamond City.) When alerted, a six-man crew—four oarsmen, a helmsman, and a harpooner—launched a skiff in chase of their quarry.

Northern right whales, so named because they swam slowly, floated when dead, and yielded considerable amounts of oil, were the primary targets.[1] They also migrated near the shore on an annual trek between their winter calving grounds off Georgia and northern Florida and their summer foraging areas in the Gulf of Maine and the Bay of Fundy. A successful hunt ended when the carcass was towed to shore and its blubber removed and boiled in fifty-gallon iron pots, known as try kettles. Rendering, or "trying," completed, the oil was ladled into barrels for storage and later used as a fuel or lubricant. The mouths of right whales also provided sheets of baleen, a flexible, fingernail-like material that filtered food from the water and from which combs, buggy whips, and corset stays were manufactured. Sperm whales, while less frequently encountered near the shore, were a special prize: they yielded not only blubber for rendering but also a high-quality waxy oil (spermaceti, contained in their heads), used to produce smoke-free candles, and ambergris (a natural by-product found in their digestive tracts), used in perfumes. Instead of baleen, the lower jaws (only) of sperm

1   Overhunting had extirpated the Atlantic population of gray whales early in the Colonial era. Like right whales, they migrated near the shore on a predictable schedule and hence were vulnerable to land-based whaling operations. Their skeletons occasionally turn up on the Outer Banks. Gray whales persisted in the Pacific Ocean despite heavy exploitation, but thanks to stringent protection since 1949, they have returned to their former numbers.

whales bear large, conical teeth, on which clever whalers carved images—a form of folk art known as scrimshaw.

North Carolina whalers sometimes named the whales they captured, usually because of an association of some sort; the practice was workable because usually fewer than five whales were killed per year. George Washington Whale, for example, was landed on the president's birthday, and Cold Sunday was killed on an unusually frigid day. The skeleton of another, a particularly ferocious right whale dubbed Mayflower, is on display at the North Carolina Museum of Natural Sciences. Shore-based whaling in North Carolina ended in 1916, although ships from New England continued whaling off Cape Hatteras for another decade.

A related industry emerged early in the eighteenth century when bottlenose dolphins were netted near Beaufort and on the shores of Hatteras Island. The long seines—handled in two-hundred-yard sections—captured four hundred to five hundred dolphins each year; the nets formed a semicircle around the dolphins, which were then hauled ashore. (Some of the trapped animals were shot from boats as well.) Five or six animals yielded one barrel of oil, which provided fuel for the lanterns of local lighthouses. Dolphin hunting ended with the beginning of the Civil War.

A final word about northern right whales: with fewer than four hundred remaining, they face a grave future. Although protected from commercial hunting since 1935, their migratory route along the coast corresponds with major shipping lanes, which has resulted in collisions between the slow-swimming whales and hefty vessels. Speed restrictions now require that ships move at reduced speeds when sailing in certain locations at specified times—a measure that has lessened but not eliminated the fatalities. Entanglements with fishing gear, however, remain a serious threat and were involved with more than half the strandings of large whales (of five species) on the North Carolina coast for which human interactions could be assessed. For northern right whales, such losses have a direct bearing on their conservation: preventing the deaths of just two adult females per year will maintain the growth rate of the population at its replacement level.

Shaped by Salt: Maritime Forests

Although now much reduced in their extent, maritime forests still rim some areas along the coastline of North Carolina, most prominently on the Outer Banks. Beachfront homes and other coastal developments today interrupt what once were far larger tracts of these distinctive woodlands, leaving behind fragments as sentinels for an ecosystem at risk.

As described earlier, maritime forests typically begin in slacks between dunes where thickets of wax myrtle and yaupon—each with waxy foliage that resists salt burn—gain shelter from the harsh shorefront environment. Openings in these thickets in turn offer protected sites where a forest can develop in the next and final step of succession.

Live oak, commonly in association with laurel oak, dominates the overstory of maritime forests in North Carolina; these form a tightly knit canopy that inhibits salt spray from reaching the underlying vegetation. The durable wood of live oak, along with the desirable curvature of its limbs, attracted nineteenth-century shipwrights, who sent crews of "live oakers" from the shipyards of New England to southern coastlines in search of structural timbers. In time, the advent of steel-hulled ships ended the pillaging, but by then much of the maritime forest had already fallen to the woodsman's axe.

Loblolly pine and red cedar often occur in the overstory as well, but of these, only the latter resists salt burn. In fact, a distinctive form of the latter—coastal red cedar—is now recognized in part because of its higher tolerance of salt in comparison to those growing elsewhere, whereas loblolly pines suffer when storms or other disturbances open the canopy. Red bay, dogwood, dwarf palmetto, ironwood, and yaupon characteristically form the shrub layer of the understory, which includes tangles of greenbriar, grape, and other vines, of which poison ivy deserves mention. Red bay also occurs widely in pocosin and Carolina bay communities, but the species unfortunately faces an uncertain future throughout its range. Except for partridgeberry, few herbaceous plants can cope with the small amount of light that penetrates the tight canopy; hence, ground cover remains sparse.

Maritime forests remain unique not for their floral composition (the same species occur widely in other settings) but because they form a community shaped by salt spray. This powerful influence imports calcium, potassium, and other metallic cations into the nutrient-poor quartz sands prevailing at these sites and sculpts the defining feature of maritime forests: a sloped canopy. In the past and even today, the canopy's wedge-shaped contour was credited to the relentless force of onshore winds, but it actually forms from the harmful effects of salt carried ashore by these winds, a discovery of ecologist B. W. Wells. The spray, technically an aerosol, repeatedly kills successive generations of terminal buds on the seaward side of the canopy, whereas more buds survive and continue growing as the distance from the surf increases. The difference in growth thus creates a distinctive slope (Fig. 2-9). Meanwhile, the salt-resistant leaves of live oak close into a tight canopy that protects the underlying vegetation from salt damage. It follows, however, that openings in the protective shield will harm the understory vegetation, which affects its composition and alters the temperature and moisture regimes within the forest. In short, the integrity of the canopy determines the nature of the community.

Remnants of a maritime forest confirm the restless movements of dunes in the wake of storms. A series of hurricanes did just that in the late 1800s when they destroyed the plant cover holding the foredunes in place on Shackleford Banks. No longer anchored, the sand moved across the island and buried a tract of maritime forest. Decades later, the forest emerged skeleton-like as the loose sands continued their journey; most of the bones were live oak and red cedar, the latter of which better endured exposure in the years that followed (Fig. 2-10). Some liken the setting to a graveyard forest of ghost trees. An observer in 1917 estimated that the wall of sand advanced at a rate of four to twelve feet per year. These storms also initiated the emigration of the island's residents, who settled in the relative safety of Harker's Island, leaving Shackleford Banks thereafter without permanent residents.

Fortunately, blocks of maritime forests survive under the protection of federal, state, or private conservation agencies. Among these are forests

FIGURE 2-9. The searing sting of salt aerosols kills the buds on the windward edges of maritime forests and coastal thickets. The influence of the aerosols steadily diminishes toward the leeward edge, producing the distinctive sloped canopies of these communities.

at Kitty Hawk, Buxton Woods, Nags Head, Bogue Banks, and Bald Head Island. In some areas, overabundant populations of white-tailed deer may threaten the vegetation and require measures to reduce their numbers, but the curtailment of development in these preserves removes the primary threat confronting maritime forests.

FIGURE 2-10.
Stands of trees smothered by sand and later exposed formed a "ghost forest" on Shackleford Banks. Photo courtesy of the National Park Service.

A Rocky Spot on the Beach

Sandy beaches, especially those on barrier islands, characterize most of North Carolina's long shoreline. One site, however, stands alone: a rocky outcrop of coquina at Fort Fisher on the ocean side of the Cape Fear Peninsula (Fig. 2-11). These blocks of sedimentary rock formed from shells (or in some areas coral) that were cemented into a matrix during the Pleistocene; indeed, the name originated from the Spanish word for clam. The formation near Fort Fisher extends from the shoreline outward into the ocean across the continental shelf, where it acts as a natural jetty that alters the movement of sand on nearby beaches. On land, however, coquina remains buried by other strata, although the formation was exposed during the excavation (completed in 1931) of Snow's Cut, where it can be seen in places along the canal's edge. Ecologist B. W. Wells considered coquina as "the backbone of the lower [Cape Fear] peninsula," a reference to its capture of sediments that slowly fashioned the present-day landform.

The presence of coquina near Fort Fisher may be the northernmost occurrence of its kind on the eastern coast. (Coquina Beach on the Outer Banks near Kitty Hawk gained its name from coquina clams and not for a rocky formation as described here.) Coquina provided a local source of building material for Sedgeley Abbey, a Colonial-era mansion on the lower Cape Fear River. Abandoned after the Civil War, the imposing building was demolished, and the phosphate-rich stones were burned and applied as fertilizer. In Florida, where coquina occurs more commonly, Spaniards constructed forts of the soft rocks because the walls could absorb the impact of cannonballs without crumbling.

Fossils of at least eight species of clams and two of snails have been identified in the coquina matrix at Fort Fisher, and most of these resemble modern forms. A million or so years later, the site still maintains an organic role, in this case as habitat for modern marine life. Species such as blue mussels and oysters attach themselves to the hard surface and in turn attract their predators. At high tide, seawater fills low spots between the rocks that persist as isolated pools when the tide recedes. Known as tide pools, these become microcosms alive with organisms typical of the

FIGURE 2-11. An outcrop of coquina, a conglomerate of fossil shells, at Kure Beach near Fort Fisher State Park forms the only natural rocky site on the coast of North Carolina. Small tidal pools in the formation harbor a distinctive community of marine life.

intertidal zone, among them sea anemones, chitons, starfish, limpets, crustaceans, and periwinkles, along with sea lettuce and other algae. At least one species of goby, most likely the darter goby, may also flit about in this menagerie. In such a cafeteria, shorebirds and oystercatchers seek prey confined at low tide. The rhythmic changes make life tough in these environments, not the least of which is exposure at low tide to both predators and desiccation, soon confounded by the rigors of a pounding surf as the tide returns. Hardships aside, these natural aquariums remain one of nature's more delightful treasures. With her engaging style, Rachel Carson aptly noted that tidal pools "have many moods [in which] the beauty of the sea is subtly suggested and portrayed in miniature."

Rocky shores are not without scientific importance. Those along the rocky Pacific coastline became the outdoor laboratory for Robert Paine's seminal studies of relationships within food chains. More specifically, his studies revealed the top-down importance of predators: their removal led to profound changes in the structure of the entire community, a process now formally known as a trophic cascade. With the removal of the apex predator—referred to as a keystone species—the numbers of some of the remaining organisms skyrocket, whereas others disappear entirely, leaving the community with a net loss of biodiversity.

Jockey's Ridge: A Dune to Remember

The dune systems on the Outer Banks reach their zenith at Jockey's Ridge, an immense mile-long pile of sand that may extend upward for a hundred feet (Fig. 2-12).[1] Its actual height at a given time depends on the strength of the current wind regime, but these variations aside, the dune is the largest of its kind on the Atlantic coast of North America. Geologists tag this and other large isolated dunes that move about as "medanos," but with charming imagination, locals identify the large

1   According to legend—no one really knows for sure—the dune once served as a natural grandstand where viewers watched riders race their horses on the firmer sand at its base, hence establishing "jockey" as a descriptor for the site. Another suggests that the name is simply a corruption of "Jackey," a former owner of the property.

dunes in the area as "whaleheads," almost certainly a reference to the massive front end of sperm whales. Jockey's Ridge continually moves back and forth about ten to eighteen feet per year in keeping with the seasonal changes in wind direction—from the northeast in winter and from the southwest in summer. (In 1903, the same summer winds famously helped the Wright brothers launch their historic flight at nearby Kill Devil Hills.) The huge dune consumed an ill-placed hotel in 1850 and, more than a century later, also similarly covered a miniature golf course, which at this point no longer needed a sand trap. The dune also shields the adjacent maritime forest from the full force of storms and windblown salt spray (see below).

Jockey's Ridge stands as a remnant of a long strand of large dunes that once extended into southern Virginia. Just why and how this formation vanished is unclear, but intense development surely contributed to its disappearance—a fate that nearly claimed Jockey's Ridge as well (see below). The quartz sands at Jockey's Ridge originated from eroding mountains well inland, which then were carried to the sea some three thousand or four thousand years ago by the Roanoke River; at the time, the river flowed directly into the ocean, but rising sea levels later drowned its mouth and formed present-day Albemarle Sound. The sands accumulated in offshore shoals that storms pushed ashore to build the huge dune we see today. A fine powderlike outer layer of sand overlies a denser and wetter inner core maintained by rainfall. Lightning strikes sometimes form hollow glassy tubes of heat-fused sand known as fulgurites on the dune's surface.

Unlike most neighboring dunes on the Outer Banks, Jockey's Ridge generally lacks vegetation, but what exists represents a rare type known as the Live Dune Barrens community of grasses, vines, and shrubby clumps of wax myrtle and black cherry. Predictably, the sparsely vegetated community is more affected by shifting sand than by salt spray. One of the grasses at Jockey's Ridge, purple sandgrass, while not itself rare, has evolved an unusual adaptation to reproduce on unstable dunes. Like most grasses, this species relies on wind to pollinate spikelets of tiny flowers forming an inflorescence at the end of each stem, but it also bears self-fertilizing flowers on spikelets growing within the leaf sheaths

FIGURE 2-12. Unlike most other dunes on the Outer Banks, Jockey's Ridge features scattered patches of sparse vegetation and large areas of exposed, highly mobile sand.

at nodes on the stems. Further, the stems of purple sandgrass break at their nodes and independently disperse viable seeds as the fragments tumble across the dunes. Seeds from either source—terminal inflorescences or leaf sheaths—successfully germinate even when buried nearly two and a half inches deep.

With sufficient rainfall, ephemeral pools form in depressions at the base of the dune, where they provide short-lived breeding habitat for amphibians, including squirrel treefrogs and eastern narrowmouthed toads, whose features are unlike either a typical toad or frog. These species breed in response to the availability of surface water; hence, they may begin courtship whenever heavy rains fall during the warmer months of the year. The dune itself lacks permanent residents, in part because the shifting sands preclude denning but also because the surface heats up well beyond the ambient temperature by twenty degrees or more during the summer months. As a result, animals adapted to temperate-zone conditions cannot cope with the localized environment on Jockey's Ridge.

Ghost tiger beetles, however, fare well in the harsh dune environment, but the population at Jockey's Ridge is the only one known in North Carolina. The creamy-white color of these beetles provides an almost perfect match with the sand, leading some observers to remark that it's easier to see their shadows than the beetles themselves. The adults run down their prey in short bursts: they run so fast that they often have to stop at intervals to refocus on their fleeing victims. Their grublike larvae dwell in burrows, in which they anchor themselves with two pairs of body hooks; they wait with open jaws in the entrance to ambush prey, which they grab and pull into their burrows to consume. They then throw the remains out of the burrows away from the entrance. Because ghost tiger beetles are habitat specialists (dunes, in this case), their coexistence with sandboarding and other human activities on Jockey's Ridge seems precarious.

Development long ago claimed much of the dune systems on the Outer Banks, and a similar fate almost destroyed Jockey's Ridge. In 1973, a bulldozer began eating away at the dune, but the operator wisely backed down when he was confronted by a determined woman named

Carolista Fletcher Baum. Blessed with a magnetic personality, she immediately organized the People to Preserve Jockey's Ridge and acquired enough members to gain the attention of politicians, with the result that the site became a National Natural Landmark in 1974 and a state park a year after that—by any measure, a significant achievement in just two years. Jockey's Ridge thus survived the creep of development, and for this Carolista Baum properly joins the ranks of women who likewise saved natural treasures, among them Marjory Stoneman Douglas (the Everglades) and Rosalie Edge (Hawk Mountain Sanctuary).

Forces of Nature: Hurricanes and Nor'easters

Storms shape coastal shorelines, perhaps more so on the barrier beaches of North Carolina than anywhere else on the Atlantic coast of North America. Tropical cyclones forming in the Atlantic Ocean, better known as hurricanes, come first to mind, closely followed by winter storms popularly called nor'easters—both movers, if not shakers, of inlets, beachfronts, and dunes. Moreover, in an era of rising sea levels, these storms may accelerate the relentless landward migration of barrier beaches. Because of warming ocean temperatures, hurricanes also seem likely to increase their intensity but not necessarily their frequency in the years ahead.

Hurricanes develop where warm surface waters increase evaporation rates, followed by the convection-driven upward movements of the moist air. These unstable conditions create the initial phase of what may become a hurricane—a low-pressure area known as a tropical depression. Thunderstorms form as the moist air condenses into clouds, and the updrafts continue to strengthen as more air is pulled in from the periphery of the depression. As a result, the clouds move even higher and begin rotating in a counterclockwise motion. Some tropical depressions develop no further and eventually dissipate, whereas others become tropical storms—and now worthy of a name—when their winds exceed thirty-nine miles per hour. With continued development, a tropical storm becomes a hurricane when its surface winds exceed seventy-four mph and circulate around a cloud-free eye. The Saffir-Simpson hurricane wind scale, now widely used by both meteorologists and the public, separates hurricanes into five categories, the weakest (1) with wind speeds of 74–95 mph and the strongest (5) with winds exceeding 157 mph. Another index, based on the area covered by the storm, considers the destructive capability of a hurricane, although it appears less often in pre- and poststorm headlines. Landfall occurs when the eye reaches the shoreline, although extended outer bands of wind and rain arrive much sooner. Storm surges, made worse when a hurricane hits at high tide, commonly inflict excessive destruction in addition to wind damage. Thunderstorms and tornadoes often accompany hurricanes and may strike well inland from the coastline.

Nor'easters, named for the direction from which their winds blow, usually develop between October and April, with most occurring in February.[2] They originate when a low-pressure system, often moving eastward from the continent's interior, meets Arctic air in the jet stream, then run into moist warm air from the Caribbean Sea or the Gulf of Mexico. This clash occurs along the mid-Atlantic coast, often in North Carolina, where it becomes a nor'easter, which is fueled by the sharp temperature gradient between the warm and cold air masses. The storms circulate in a counterclockwise movement as they move northward along the Atlantic coast to New England and beyond, but their winds seldom reach hurricane strength. However, nor'easters may linger for several days and accordingly drop large quantities of rain or snow during the course of their slow movement; some of these snowy events qualify as blizzards. These effects may be enhanced when a storm encounters the clockwise circulation of a high-pressure system, which can then hem in the nor'easter against the shoreline. Nor'easters often

2   Benjamin Franklin (1706–1790), among his other scientific endeavors, also took notice of nor'easters, specifically to reveal that these winter storms moved northward even though their winds blew from the opposite direction. Previously, the consensus held that storms traveled in the same direction as their surface winds; hence, a nor'easter supposedly originated in the northeast and moved south. Franklin recognized this error when he and one of his brothers separately watched a lunar eclipse from Philadelphia and Boston, respectively, in November 1743. Ben's view of the eclipse, however, was foiled by overcast skies and rain accompanied by snappy winds. He assumed his brother had experienced the same limitations. Not so—his brother had watched the eclipse in Boston without difficulty, noting that stormy weather had appeared somewhat later; ergo, the stormed had moved north, not south, along the East Coast.

produce larger waves than those of hurricanes, and their storm surges may cause greater harm because they extend over several tidal cycles. Moreover, hurricanes usually damage relatively short stretches of coastline—sixty to a hundred miles—whereas nor'easters may devastate hundreds of miles of the Atlantic coast.

For centuries, these storms have wreaked havoc on buildings, shipping, roads and railroads, crops, and communication systems, topped off by the loss of many lives. Other costs become more difficult to assess, such as beach and dune erosion, uprooted seagrasses, overflows or spills of sewage and toxic materials, or changes in salinity or currents wrought by newly cut inlets. A nor'easter known as the Ash Wednesday Storm (1962), one of the ten most destructive storms of the twentieth century, tossed thirty-foot waves against the Outer Banks. In 1989, Hurricane Hugo tore through Francis Marion National Forest in South Carolina, destroying 87 percent of the trees serving as nesting sites for red-cockaded woodpeckers and killing 63 percent of the birds. In response, biologists successfully installed artificial nesting cavities to replace those lost in the downed trees, and the forest again supports one of the largest populations of the endangered birds. Hurricane Florence (2018) produced widespread fish kills, including those in several large rivers (e.g., Cape Fear, Neuse, Chowan, and Roanoke). These kills occur when the level of dissolved oxygen, normally five to six parts per million (ppm), drops to two ppm—the result of flooding that flushes large amounts of stagnant swamp water into rivers or other bodies of water. Such waters are comparatively warm and thus less capable of holding dissolved oxygen. Hurricanes commonly destroy sea turtle nests and sometimes drown feral horses on the Outer Banks. At times, the storms create barren sandy areas where least terns and other birds establish nesting colonies.

Beacons, Birds, and Bureaus

As one of its first acts, Congress placed twelve existing lighthouses under federal jurisdiction in 1789. What later became the Lighthouse Service also constructed new structures, one of the first being a lighthouse authorized in 1792 to stand at the mouth of the Cape Fear River. Eventually,

FIGURE 2-13. Like many other lighthouses, the iconic light at Cape Hatteras played a role in the formation of what is today the U.S. Fish and Wildlife Service. As shown here, the Hatteras Light stands at its original position before beach erosion necessitated its relocation farther inland.

the beams from seven "candles in the wind" would mark the shoreline of North Carolina's Coastal Plain. By 1872 331 lighthouses guided shipping along the Atlantic coastline, with the Hatteras Light the tallest of all (Fig. 2-13).

Lighthouses warned mariners but attracted birds, often fatally. John James Audubon, in his *Ornithological Biography*, was perhaps the first naturalist to comment on this problem, noting that Canada geese "fly against beacons and lighthouses [when] attracted by the light of these buildings and [at times] a whole flock" may be victims. Henry David Thoreau also described birds falling to the ground "with their necks broken" after striking the lighthouse on Cape Cod. More than a century later, X-rays have revealed that a bird killed in a collision with a window actually dies from a fractured skull, not a broken neck.

In 1877, a young college student at Yale recognized that these collisions were more than an unfortunate loss of wildlife—they were also a unique source of biological information. The dead birds in fact were readily obtained samples that could be easily identified in hand at specific times and places, which presented a way to accurately describe bird migrations, even for those species that migrated at night. C. Hart Merriam thus wrote that "few people . . . have any idea that hundreds of thousands of birds are killed each year, during migrations, by flying against lighthouse towers" and that "nearly all our common, and many rare, migrants are found among the dead." Merriam's long paper about the birds of Connecticut, in part based on birds killed at a lighthouse marking the entrance to New Haven Bay, firmly established his place as a serious ornithologist.

Merriam went on to medical school but gave up his successful practice to pursue his passion for natural history. In 1883, he was among the founders of the American Ornithologists' Union (AOU), which elected him not only its secretary but also chairman of its Committee on Bird Migration. Merriam again sought the help of lighthouse keepers as a prime source of data.

North Carolina's lighthouses provided a fair share of samples. As examples, the records from Hatteras Light include sixty to seventy-five snipe "frequently found dead at one time," as well as teal, "sometimes as many as 8 at a time." Yellow-rumped warblers, however, topped the list, with hundreds sometimes killed in a single night. On one occasion 350 dead warblers littered the balcony around the lantern house, and another 140 fell to the ground below. Yellow-rumped warblers also dominated the fatalities at Cape Lookout, and the light at Bodie Island killed a flock of geese less than thirty days after it went into service and continued to claim waterfowl of several species. At Currituck Beach, the keeper reported that as many as five hundred bobolinks died in a single night along with lesser numbers of willets and other shorebirds; he also requested bigger packages, because "some of the birds were very large." Meanwhile, similar records—by the hundreds—poured in from lighthouses elsewhere, and Merriam quickly realized he needed help.

Big, Red, and Good to Eat

Red drum, also known as redfish or puppy drum, provides anglers with a good fight and an even better meal. However, in the 1980s, after New Orleans's chef Paul Prudhomme crafted his Cajun-style recipe for blackened redfish, red drum populations faced dire consequences. An overnight hit, the tasty dish created a demand that exceeded the supply. Commercial and recreational anglers alike overharvested red drum populations across much of their range, which extends along the coastline from southern New Jersey to northern Mexico. In response, the harvest of adults, mainly those five or more years old, was prohibited in federal waters, and each of the coastal states placed restrictive limits on the size and number of red drum that could be legally harvested from inshore waters. North Carolina, among others, manages red drum using "slot" regulations that establish upper and lower size limits. Only those fish with lengths falling within the slot may be legally harvested; all others must be released. Texas, Florida, and South Carolina also initiated hatchery programs in the late 1980s to help replenish the depleted populations.

Well named for their pinkish to copper coloration, red drum also sport one and sometimes two prominent ocelli, or "eye spots," near the base of their tails; these presumably redirect attacks from predators to a less critical area. In late summer and early fall, male red drum gather in schools at inlets and the mouths of rivers to await the arrival of females. Males mature in about three years; females in five. Males attract mates with croaking sounds produced by vibrating specialized muscles known as sonic muscle fibers against their swim bladders, which act as resonating chambers—in effect, as drums. Most spawning occurs in the nearshore ocean within sight of land. Females come and go at two-to-seven-day intervals during this period and spawn two hundred thousand to two million eggs on each occasion. Oil droplets in the fertilized eggs provide floatation, as well as nourishment for the developing larvae; the eggs hatch in about two days.

Currents carry the developing larvae landward, where they enter estuaries through inlets and settle near seagrass beds or oyster reefs. The fry grow rapidly, reaching lengths of nearly a foot within a year. The young fish avoid predators by remaining in shallow water, where they forage on crabs, shrimp, and small fishes. Red drum feed head down

in the thick vegetation, often breaking the surface with their tail fins—this behavior, known as tailing, provides a welcome calling card for anglers fishing in estuaries and sounds. At maturity, the fish leave the shallow sounds and move offshore for the remainder of their lives. In winter, older (thirty to fifty or more years of age) and much larger (typically thirty-five to forty-five inches long) individuals of both sexes, called bull reds, become highly prized targets for surf-fishing anglers. Indeed, the current world record for red drum—a fifty-seven-inch ninety-four-pounder—was caught on Hatteras Island in 1984. Today, with the upper slot limit set at twenty-seven inches, such a trophy would not be legal to keep, so carry a camera, tape measure, and scales to assure bragging rights.

In 1971, the North Carolina legislature designated red drum as the state's official saltwater fish. Texas similarly recognized the species in 2011, in part because its recovery from overexploitation serves as an exemplary environmental success story.

Merriam advised the AOU about the overload, noting that this work could not continue without assistance. In response, he was authorized to ask Congress to create the Division of Economic Ornithology, which was approved by the Senate in 1885 along with an appropriation of $5,000. The House, however, insisted that the work be carried out in the Division of Entomology. To no one's surprise, Merriam was appointed to direct the project, which focused, at least superficially, on birds that fed on harmful insects—perhaps slyly, he continued to rely on lighthouse keepers for information regarding migratory birds. Merriam also developed a growing interest in mammals and, armed with a $10,000 appropriation, a new and independent agency emerged in 1886, the Division of Economic Ornithology and Mammalogy, which included the control of rodents and predators, as well as surveys of mammals in thirty states and northern Mexico. Merriam would remain in charge until he resigned in 1910.

After yet another name change, the agency became the Bureau of Biological Survey in 1905, and it remained unchanged until 1940, when the bureau merged with the U.S. Fish Commission to become the U.S. Fish and Wildlife Service, housed in the Department of the Interior. Today, the Fish and Wildlife Service oversees federal wildlife refuges and fish hatcheries, endangered species, waterfowl and other migratory birds, and a host of management and regulatory activities funded by a budget of more than $1 billion—a mission that began when lighthouse keepers swamped a young ornithologist with their lists of dead birds.

Mustangs on a Sandy Shore

One can pick and choose among the ideas about the origins of the horses on the Outer Banks (Fig. 2-14). Commonly regarded as "wild," these horses ("ponies" to some) are more accurately described as feral—untamed animals descending from domesticated ancestors. The most popular—and certainly the most romantic—notion proposes that the stock originated as castaways from Spanish galleons wrecked in the sixteenth century. Ships returning to Spain sailed north along the North Carolina coastline to take advantage of the Gulf Stream, but they also encountered treacherous sandbanks in the uncharted waters of what later became known as the Graveyard of the Atlantic. Critics, however, point out that avaricious Spaniards would hardly devote cargo space to horses and their fodder at the expense of carrying additional treasure homeward. Moreover, why carry horses back to Europe, where they already flourished, when it was simply more expedient to leave them in the New World ports from which the Spaniards sailed? Others doubt that any horses that might have struggled ashore did so in numbers large enough to establish viable populations, thereby falling short of a critical mass and perhaps also wanting for a favorable mix of sexes. Another theory proposes that the horses descended from stock that colonists brought to the Outer Banks. Fencing was unnecessary on the islands; hence, the free-roaming herds were simply abandoned when the colonies failed. Whatever the explanation might be, nearly everyone agrees that horses have lived on the Outer Banks for at least three hundred years. And for residents and visitors alike, the sturdy mustangs emerged as a living resource and national treasure.

During their long tenure and isolation on the Outer Banks, the horses adapted to local conditions. Marsh grasses provide half of their year-round diet, with much of the balance coming from sea oats and other dune vegetation. Black needlerush, although abundant, remains untouched, no doubt because of its coarse and probably indigestible tissues. Foliage in maritime forests represents less than 5 percent of forage consumed by the horses—too little to alter either the structure or the composition of these woodland communities. In the marshes, however, grazing may remove a considerable amount of vegetation, although the damage may be repaired. Because the horses consume only the aboveground parts of the marsh grasses, the plants' vital rhizomes and root systems remain intact and able to regenerate their stems and leaves, aided by the steady influx of nutrients with each tidal cycle. Moreover, because the dominant species—smooth cordgrass—lacks invasive competitors, heavy grazing does not change the composition of the marsh vegetation, as often occurs in overgrazed terrestrial communities.

Still, any plant community can suffer serious or even irreparable damage when too many animals consume vegetation in amounts beyond its ability to recover—in other words, when grazers exceed the carrying

FIGURE 2-14. Although more accurately identified as "feral," wild horses have a long history on the Outer Banks, where they have adapted to the harsh conditions of a coastal environment. Along the way, they have also captured the admiration and affection of residents and visitors alike. Photo courtesy of the National Park Service.

capacity of their habitat. Protecting dune vegetation highlights these fears, and trampling presents a potential threat to the nesting activities of sea turtles and shorebirds. The size of the population presents a dilemma for other reasons as well. A large number of horses increases genetic diversity (normally a benefit), but it also risks the loss of the very traits that make the population unique. Conversely, too few horses increases inbreeding, which in turn promotes the emergence of undesirable traits that may jeopardize the population, among them physical abnormalities and sterility. In all, the management of relatively small insular populations must deal with both environmental and genetic concerns.

With few sources of permanent fresh water available, the horses learned to paw pits in search of drinking water. The pits, two to four feet in depth, tap a supply of fresh water contained in an inverted dome "floating" atop a heavier aquifer of salt water. This phenomenon, formally known as the Ghyben-Herzberg principle after its Dutch discoverers, commonly occurs on islands surrounded by salt water. The pits may remain in use for some time, but they eventually fill in or become fouled, forcing the horses to create new excavations. This vital adaptation evolved not only out of necessity but also in concert with local conditions—sandy soils that facilitated excavation by hooved animals.

The horses survive storms, including hurricanes, rather well, but drownings sometimes occur. No horses were lost in the rage of Hurricane Florence (2018), for example, but twenty-eight drowned when a storm surge likened to a "minitsunami" swept across Cedar Island during Hurricane Dorian (2019). For the most part, the horses simply "hunker down" on higher ground, often in the shelter of thickets and maritime forests, but dominant stallions at times may force subordinate horses to seek shelter elsewhere, which may lessen their chances of survival. Lacking shelter, harems may merge and together turn their rumps—"croups" in horsespeak—into the wind for protection. Storm surges also may fill the excavated pits with salt water and thus add another stress to an already difficult situation. All told, however, the furies of coastal storms have surprisingly little effect on the doughty horses despite living in a tempest-prone environment.

The seeds of what would later lead to the "Pony Wars" were sowed in 1966 when segments of the Outer Banks, most notably Shackleford Bands, were designated as a national seashore and a unit of the National Park Service (NPS). Under its mandate, the NPS frowned on nonnative wildlife, especially livestock, which thus were to be removed one way or another. Predictably, that decision was vigorously opposed by those who had for generations taken a passionate and proprietary interest in the horses as part of their cultural heritage. In response to the outcry, the NPS removed the goats, sheep, and cows from Shackleford Banks but left the horse population intact. Nature took over, and while the census overestimated their numbers, the horses essentially doubled between 1987 and 1994, leading the NPS to remove 120 for adoption. The remaining population was treated with contraceptives, but in the process, several horses testing positive for equine infectious anemia (EIA) were euthanized.

EIA, an incurable viral disease, affects only horses but shares some similarities with HIV in humans. For the most part, biting flies transmit the disease, but it can also spread from horse to horse by contact with blood or other bodily fluids. Some horses, although infected, remain healthy but harbor the disease, thus serving as reservoirs for more infections. The conflict between the residents and the NPS heightened in 1996 when seventy-six horses infected with EIA were euthanized before a newly formed citizens' group, the Foundation for Shackleford Horses, could locate quarantine sites. The balance of the herd, 108 horses, tested negative and remained on the island. These events marked the onset of the Pony Wars, a two-year conflict about the future of the Shackleford horses.

Roundups, contraception treatments, and EIA testing continued thereafter, but the issue was not fully resolved until 1998, when the Shackleford Banks Wild Horse Protection Act became law. The legislation recognized that a minimum of 100 horses were needed to maintain the population's genetic viability and that the population should not exceed 130 in order to thwart the harmful effects of overgrazing; excess animals were placed for public adoption. Moreover, the law mandated that the NPS and the foundation were henceforth to comanage the herd—a partnership that still prevails with the careful stewardship and the good intentions of both groups. In 2008, North Carolina recognized the breed as the official state horse.[3]

So what about the origins of the horses? Analyses of their genetic makeup revealed a marker known as Q-ac, which characterizes horses with a colonial Spanish ancestry. Not overwhelming evidence of ship-wrecked survivors, to be sure, but certainly enough to keep the legend alive and well for generations to come.

Sea Turtles: Ancient Reptiles on a Modern Shore

Seven species of sea turtle ply the world's oceans, each battling to survive. Five species nest on the coast of North Carolina, mostly loggerheads but much less frequently also green sea turtles, leatherbacks, Kemp's ridleys, and hawkbills (Fig. 2-15). The number of nests per year varies considerably, ranging from 799 to 2,357 (an average of 1,465) during a five-year period (2015–2019). Loggerheads nesting in North Carolina, along with South Carolina and Georgia, form a "northern recovery unit," which recognizes their genetic difference from those nesting in Florida. The beaches at National Seashore units such as Cape Hatteras and Cape Lookout provide about 40 percent of the nesting habitat for sea turtles in North Carolina; respectively, 472 and 528 nests occurred at these sites in 2019. Given warm water as late as November or even December, the extensive estuarine habitat in North Carolina also provides prime foraging areas for juvenile sea turtles hatching in Florida, the Gulf of Mexico, and the northern Caribbean. Green sea turtles seek seagrasses, whereas loggerheads and Kemp's ridleys feed on the abundance of marine invertebrates in these environments.

Loggerheads may weigh well over four hundred pounds and grow to lengths of nearly forty inches, as measured by the length of their upper shells (carapaces). As adults, females come ashore only to nest; otherwise, like males, they remain at sea. Loggerheads belong to a family that

3 "Breed" as loosely used here includes all of the feral horses on the Outer Banks, but, according to equine geneticist E. G. Cothran, only those on Currituck Banks at Corolla effectively represent a "breed unto themselves." Other herds occur on Hatteras Island and at the Rachel Carson North Carolina National Research Reserve.

FIGURE 2-15. More than a thousand loggerhead sea turtles nest each year on the sandy shores of the North Carolina coast. Federal, state, and private conservation organizations closely monitor the fate of their nests. Photo courtesy of the National Park Service.

appeared at least forty million years ago, although their ancestors likely evolved even earlier.

Loggerheads, named for their large heads, consume a variety of foods, ranging from bottom-dwelling invertebrates (mollusks, sponges, starfish, and sea urchins, among many others) to algae and vascular plants. Unfortunately, they also eat jellyfish, which may turn out to be a fatal meal of plastic bags. Younger loggerheads manipulate their food using pointed scales on the forward edge of their forelimbs; these "pseudo-claws" rip apart larger chunks held in their powerful jaws.

When mature, female loggerheads mate at sea and return to nest on or near the same beach where they hatched thirty or more years earlier—a textbook example of homing behavior formally known as natal philopatry. A once-popular theory, now rejected, held that hatchlings were imprinted with chemical cues they acquired at the nesting area and that years later these cues guided the same individuals as adults back to the same site. Although speculative, a newer theory proposes that this ability at least in part seems associated with the Earth's magnetic field; in other words, loggerheads and other sea turtles can navigate by "reading" a map of geomagnetic coordinates. Intriguing, to be sure, but as matters stand, the turtles have yet to reveal their secret but effective navigational system.

Gravid females come ashore, usually at night, leaving behind a line of distinctive tracks (or "crawls") en route to a site above the high-tide line, where they dig net cavities in which to lay and then cover about 100 to 120 eggs before returning to the sea. During a single nesting season each female will repeat this process several times—usually two to three, but sometimes up to eight—at about two-week intervals, then skip two to five years before breeding again. During incubation, nest temperatures determine the sex of the hatchlings; at ninety degrees most become females, whereas at eighty-two degrees most become males. (Nest temperatures generally range between seventy-nine and ninety degrees; hence, broods normally contain some of each sex.) The hatchlings emerge about two months later in an event called a "boil," then scramble toward the surf, guided by the brighter horizon over the ocean and away from elevated silhouettes of dunes or shoreline vegetation.

Life can be hard for loggerheads, beginning with egg-eating predators such as raccoons, red foxes, and, more recently, coyotes, which may destroy an entire clutch. For hatchlings, the trek to the surf may be fraught with additional predation, commonly from crabs and gulls. Once in the water, the hatchlings swim nonstop for the first forty-eight to seventy-two hours to deeper waters, where many find food and cover in masses of the floating seaweed *Sargassum*. Still, losses continue from a new suite of predators, including sharks, attracted by the menagerie of marine life sheltered by the rafts of *Sargassum*. Loggerhead hatchlings produced on beaches in the southeastern United States travel to the eastern Atlantic, where they remain for a decade or more. After reaching about twenty inches in shell length, they journey back to the western Atlantic, where they forage in sounds and other coastal waters—a move that may maximize growth and minimize predation during the next twelve to fifteen years. When mature, most again return to the open sea, where, except for nesting females, they stay for the remainder of their lives. In all, only a few hatchlings will reach breeding age.

The Lost Legacy of North Carolina's "Cannery Row"

Menhaden, also called mossbunker or "bug-heads" because of parasitic isopods lodged in their mouths, nonetheless are locally known as shad in coastal North Carolina. They run along the Atlantic coast between Maine and northern Florida in schools that may cover several acres. The densely packed schools travel near the surface, where they resemble an oil slick, a telltale feature well-known by fish-eating birds and commercial fishermen. Whereas the bony fish make poor table fare, oils processed from their flesh provide ingredients for products such as cosmetics, paint, and soap, and they replaced whale oil as fuel for lamps. When pulverized, the carcasses also produce fertilizers and high-protein meal for poultry and livestock, as well as for fish raised in aquafarms. Today, menhaden provide the principal component for omega-3 dietary supplements. Because of their multiple uses, some refer to the species as the "soybean of the sea." Menhaden also form a dietary staple in the diets of several game fishes.

A member of the herring family, menhaden spawn at sea between March and May. After hatching, the fry move to estuaries and later form schools with those from other areas as they migrate south for the winter, then return north the following spring; the schools consist of individuals of the same age/size class. In appearance, menhaden are laterally compressed, silvery, distinctively marked with a black "shoulder" spot behind each gill, and propelled by deeply forked tails. Lacking teeth, they instead filter-feed, first on phytoplankton but later on zooplankton. As adults, menhaden reach an average length of about twelve inches.

Native Americans fertilized their crops with menhaden, a practice adopted by the European colonists who settled in New England. In time, however, the use of whole fish gave way to processed fertilizer, which quickly replaced the costly guano imported from Peru. Moreover, after it was discovered that oil could be produced by boiling the fish, numerous processing factories began operations along the northeast coast. By the 1880s, however, overharvesting had collapsed the menhaden fishery in New England, and the industry moved south to Beaufort, North Carolina, and other locations where the descendants of slaves presented an able and willing labor force. By this time, purse seines had greatly improved the catch; these were fifteen-hundred-foot-long nets deployed by sixteen-to-twenty-man crews in each of two thirty-foot skiffs that encircled the "slicks," one advancing from each side. When they joined, the crews dropped a three-hundred-pound weight called a tom, which pulled together the bottom of the net like a drawstring to form a "purse" in which tons of menhaden were trapped. Now the real work began.

To bring the heavy load to the surface where the fish could be loaded onto the mother ship, the crews hauled the net upward hand over hand—grueling, backbreaking work that required the endurance of strong men. A catch might take more than an hour to "rise," and chanteys synchronized the tug and pull against the weight of the haul. Chanteys originated as African traditions that arrived with slaves and persisted on plantations. In practice, a leader provided the first line, followed by a response from the full crew singing in close harmony. Many of the chanteys reflected local situations, some were about the ship's captain (and not always favorably), others were about the women waiting onshore, still others were religious, and some were much less savory, but all provided a rhythm for hard work.

In the early 1960s, hydraulic power blocks replaced muscle as the means for "raising fish," and as the crews melted away, so did the chanteys. The menhaden fishery itself would soon face extinction, despite the continuing demand for fish meal. Environmentalism was in vogue, and the stench from processing fish became an issue in which the "smell of money" no longer fit local economies.

Beaufort, in fact, had promoted itself as North Carolina's Crystal Coast to attract tourists and retirees, a transformation bolstered by a waterfront renewed with gift shops, restaurants, and marinas, none of which were compatible with offensive odors or the Rust Belt appearance of the aging ships and factories. Beaufort's version of Steinbeck's Cannery Row, once an economic force, had become a liability. Moreover, anglers joined the fray, believing that removal of so many bait fish was "breaking the food chain," which ended with king mackerel, striped bass, and other game fish. Thus embattled, the last processing plant in Beaufort closed in 2005.

In 1988, a group of retired African American fishermen at Beaufort formed the Menhaden Chanteymen. They revived the chanteys in public performances, including those at Carnegie Hall and the National Council on the Arts and on national television and radio.[1] Before disbanding, they recorded a selection of their chanteys as *Won't You Help Me to Raise 'Em*, although, sadly, the CD is no longer available. Thus ended a cultural heritage initiated by hardy men who once pulled nets of glistening fish up from the sea.

1   Another plant at Reedville, Virginia, the last of its kind on the eastern seaboard, still processes menhaden. As they did at Beaufort, retired African American fishermen there also formed a group— the Northern Neck Chantey Singers—that performed and recorded their chanteys, but they too are no longer active, nor are their recordings still available.

Humans directly or indirectly harm sea turtles. Artificial lighting on shore, for example, may disrupt the hatchlings' journey to water, and larger turtles too often drown in fishing gear. Because of the latter, shrimp trawlers must now fish with nets equipped with turtle-excluding devices—in essence, trap doors that allow sea turtles to escape—but others drown in gillnets or get caught both on long lines set in the open ocean and on recreational fishing gear. Others suffer dire consequences when they ingest plastic litter. For hatchlings, highly skewed sex ratios favoring females seem likely to increase, given the current trend of global warming—such "mass feminization" of the population does not bode well for future production (i.e., an ever-growing proportion of females will not find mates). Meanwhile, a small army of volunteers closely monitors sea turtle nests along North Carolina's coastline, often protecting these sites with fencing to exclude predators and working with homeowners to reduce lighting near nests at hatching time. In the United States, both federal and state government agencies list loggerhead turtles as threatened, which protects this species (and others) with rigorous laws. In some countries elsewhere, however, poachers continue to take nesting turtles and their eggs for food.

Snow Geese on the Outer Banks

In 1934, Clarence Cottam, then a junior biologist with the forerunner of the U.S. Fish and Wildlife Service, completed a field trip to Pea Island and concluded that "the biological fitness of the area would make it the best waterfowl refuge along this barrier reef of islands [and] few areas along the coast would be of equal value." Thus in 1937 what is today known as Pea Island National Wildlife Refuge was established on the Outer Banks. However, in keeping with the dynamics of barrier islands, the location has been an on-again, off-again island with the opening and closing of New Inlet near Rodanthe—when the inlet closes, Pea Island becomes a northern extension of Hatteras Island. The refuge is the namesake of beach bean (or "pea"), a legume with vines that trail across sandy areas, often climbing the stems of sea oats and other coastal grasses. Mycorrhizal fungi increase the ability of this species—a relative of the commercially important soy bean—to withstand the salinities of coastal environments. When mature, the pods of beach bean forcibly eject four to eight woolly seeds that once provided a natural food source for flocks of greater snow geese that overwintered on the Outer Banks.[4] By the 1930s, however, the dwindling numbers of greater snow geese warranted the formal protection and management afforded by a federal wildlife refuge at Pea Island.

Once numbering just three thousand birds, the population of greater snow geese increased to about forty thousand by 1960, thanks to Pea Island and other refuges and, beginning in 1931, to closed hunting seasons. By 1975 the population had grown to about one hundred thousand, which allowed hunting to resume. Then things suddenly changed—the population of greater snow geese had jumped to about one million by 2008, and, even more remarkably, the numbers of lesser snow geese had mushroomed to about five million (Fig. 2-16). Indeed, the growth rate reached 9 percent, enough for the population to double every eight years. The phenomenal growth resulted from the increased availability of food during the winter months owing to grains left in fields after harvest and to the protection afforded by refuges. With such a food supplement, more snow geese survived the stresses of winter and spring migration; moreover, the well-nourished birds experienced better nesting success than ever before. Thus, mortality diminished, while production increased.

A different situation developed in the Arctic, where the swollen nesting colonies overpowered the food resources to the point that large areas of tundra wetlands turned into muddy wastelands known as eat-outs. When foraging on natural foods, snow geese "grub" for rhizomes and other underground parts of plants (as opposed to grazing aboveground,

---

4   Ornithologists recognize two subspecies based primarily on size—greater snow geese associated with eastern North America and lesser snow geese with the continent's interior. Each of the two subspecies comes with a darker color variant known as blue-phase snow geese, but this trait is far more common in lesser snow geese than in their larger relatives. (In the past, the white and blue phases were considered separate species until genetic evidence proved otherwise.) Blue-phase birds represent about 40 percent of the lesser snow geese wintering in North Carolina. Both subspecies feature a dark "grinning" patch on both sides of their pinkish bills.

FIGURE 2-16. Large flocks of snow geese once overwintered at Pea Island National Wildlife Refuge, but, as shown here, they now concentrate at Lake Mattamuskeet and other refuges farther inland.

as do Canada geese), which in effect turns a hungry bird into a living hoe. But with so many more birds seeking food, the devastation caused by grubbing becomes so severe that the slow-growing Arctic vegetation may not recover for decades and perhaps even longer. Similar damage also occurred in some salt marshes where the birds overwintered. In short, snow geese were destroying their own habitat, and a calamitous population crash seemed certain if nothing changed. For wildlife managers, the situation had become "an embarrassment of riches."

In response, a special conservation order was implemented in 1999 (1997 in Canada) expressly to reduce the populations of both subspecies by half; this allowed hunting at certain times and places previously banned by law, particularly at locations where the birds briefly stopped during spring migration. Daily bag limits for snow geese also were significantly increased, and in some states these were removed altogether. Additionally, some long-standing restrictions were waived, including those that banned the use of electronic calling devices and limited the number of shells loaded in repeating shotguns (i.e., plugs that limited the guns' capacity to three shells were no longer required). Whereas these drastic measures worked to some extent, lesser snow geese still remain far too numerous in relation to the carrying capacity at their Arctic nesting areas. In North Carolina, about ten thousand to fifteen thousand lesser snow geese currently winter at Mattamuskeet National Wildlife Refuge; others overwinter at nearby refuges (see Chapter 5). In contrast, the effort reduced the national greater snow goose population by about one-third, and while now stabilized, their numbers also remain above the desired level (i.e., a goal of 500,000 to 750,000). Today about a thousand greater snow geese spend the winter at Pea Island National Wildlife Refuge. Beach beans still trail across the sand, but unlike their ancestors, the snow geese also visit fields in search of waste grain. In addition, about sixty thousand to eighty thousand greater snow geese currently winter at Pocosin National Wildlife Refuge some sixty miles west of the refuge at Pea Island.

Pheasants in the Dunes

For hunters and birders alike, fields of grain stubble in the American Midwest epitomize prime habitat for ring-necked pheasants. Indeed, South Dakota selected the long-tailed fowl as its state bird. Pheasants traveled a long way from their native lands in eastern Asia—the species is sometimes called the Chinese ring-necked pheasant—spreading westward in cages, perhaps as items of trade. The birds eventually gained a foothold in the British Isles, likely arriving in the baggage of Roman legionnaires, thereby earning yet another geographical appellation: English ring-necked pheasant. Then in 1881 the first successful release of pheasants in the United States occurred in Oregon, later followed by releases wherever cereal grains dominated the landscape, notably including Illinois, Iowa, Nebraska, and South Dakota. The birds in fact flourished on the agricultural "grasslands" that replaced the native prairies and plains broken by the steady march of tractor and plow. They have quickly become a prized game bird, and hunters today harvest well over two million pheasants annually. However, pheasants failed in the southeastern states, presumably because of limitations posed by heat and humidity—except for a relict population persisting in highly atypical habitat on the Outer Banks.

Hunting clubs introduced pheasants on the Outer Banks between 1931 and 1935, when 175 birds were released on Hatteras Island. The birds thereafter dispersed across inlets to Ocracoke and Pea Islands (the latter at the time was a separate entity). Later, they appeared on North Core Banks, arriving either as releases or by their own means. In time, these populations winked out, or nearly so, from all but Core Banks, where hunters may still harvest a few birds each year from a population—perhaps numbering up to 350 birds—that undoubtedly fluctuates widely given the adversities of hurricanes and nor'easters. Only the colorful males may be legally included in the daily bag limit of three birds. Because pheasants have evolved a polygynous mating system, a single male may mate with eight or more females during the course of the breeding season; hence, the survival of relatively few males is enough to ensure adequate reproduction.

In the 1970s the Wildlife Commission stocked pheasants live-trapped on the Outer Banks at locations on the mainland. The effort was prompted by the hope that these birds, already acclimated to eastern North Carolina, might establish a nuclear population that would spread to other parts of the state. However, the project failed, and the Outer Banks remains the home for the only wild population of pheasants in North Carolina and, indeed, in the eastern United States. (Pheasant numbers in other eastern states remain bolstered by releases of birds raised in game farms.)

Wax myrtle fruits (berries borne only on female plants) provide the foremost food in the fall diets of pheasants on Core Banks; based on a sample of thirty birds, the gizzards of more than 90 percent contained the berries, which also represented almost 30 percent of the total volume of food. However, the inclusion of this dietary staple presents an interesting metabolic situation. The wax coating on the berries, once rendered by settlers for candles, resists digestion in all but a few birds. Among the latter, yellow-rumped warblers—once rather fittingly known as myrtle warblers—readily absorb and metabolize the wax, which largely consists of saturated long-chain fatty acids (Fig. 2-17). In fact, the warblers assimilate more than 80 percent of the fatty acids in the berries, in part because of the birds' elevated concentrations of bile salts. Whether pheasants likewise have such adaptations remains unclear. Still, the prominence of wax myrtle in the diet of pheasants presumably reflects their ability to realize some energetic benefits from the berries of this commonplace shrub of the Carolina coast.

Even yellow-rumped warblers, however, cannot maintain their body weight on an exclusive diet of wax myrtle berries, which points to the importance of other foods in the diets of both the warblers and pheasants. Indeed, the thickened underground stems—rhizomes—of pennywort rank second in the pheasant diet on the Outer Banks. To gain access to the rhizomes, the pheasants scratch out pits up to three inches deep and then clip the fleshy rhizomes into half-inch-long segments for eating. Quite likely, the rhizomes offer pheasants a succulent, protein-rich food at a time of year, fall and winter, when the birds may experience a nutritionally poor diet (i.e., seasons when fewer insects provide sources

FIGURE 2-17. In winter, the fruit of wax myrtles becomes a staple in the diet of yellow-rumped warblers, one of the few species of birds that can effectively digest the wax-coated berries. The birds switch to an insect-dominated diet in the warmer months.

of protein, although grasshoppers remain available at least until a hard freeze occurs). In any case, the effort that pheasants expend to secure the rhizomes underscores the dietary importance of pennywort.

Other foods consumed by pheasants include the seeds of beach bean (the "pea" of Pea Island) and sea rocket, each associated with primary dune systems. Sea rocket, in fact, grows along the seaward base of primary dunes, a particularly harsh environment because of its direct exposure to blowing sand and salt aerosols arising from the surf. Thus, to secure sea rocket seeds, pheasants necessarily venture down the dune face and forage along the upper edges of the beach. Moreover, because the sea rocket plants are scattered, the birds must travel at length along the beachfront to gain a crop-full of seeds. That pheasants chose to forage in such an atypical setting indicates that sea rocket represents a highly desirable food as opposed to other choices available on the flats behind the dunes.

Just why the birds survive on the Outer Banks and not on the mainland remains unclear, although several theories attempt to explain the mystery, among them fewer predators, a calcium-rich soil, and cooler seaside temperatures. Whatever the case may be, pheasants on the Outer Banks cope with an environment quite unlike the surroundings they experience in the grain belt of North America. For one, the sandy soils lack the fertility of the rich loams of the continental interior, which precludes successful farming. Further, freshwater ponds rapidly succumb to evaporation and drainage into the porous soil, and no streams flow on the islands. Salt, of course, presents a constant influence, whether from storm-driven flooding or from the surf-borne aerosols that daily waft over the narrow island. Such factors obviously govern the vegetation available as sources of food, yet the native vegetation on the Outer Banks offers pheasants a versatile grocery of native foods. Although remarkably isolated in both geography and habitat, the ring-necked pheasants that persist on the barrier islands remain a curious but welcome addition to the biota of North Carolina.

READINGS AND REFERENCES

Anatomy of the Outer Banks

Art, H. W., F. H. Bormann, and G. K. Voigt. 1974. Barrier island forest ecosystems: role of meteorologic nutrient inputs. Science 184:60–62. [Identifies salt aerosol as a primary source for calcium and other nutrients in maritime forests.]

Burns, J. M. 2015. Speciation in an insular sand dune habitat: *Atrytonopsis* (Hesperiidae: Hesperiinae)—mainly from the southwestern United States and Mexico—off the North Carolina coast. Journal of the Lepidopterists' Society 69:275-292. [Includes a formal description of a new species, the crystal skipper, and its life history.]

Cousins, M. M., J. Briggs, C. Gresham, et al. 2010. Beach vitex (*Vitex rotundifolia*): an invasive coastal species. Invasive Plant Science and Management 3:340–345.

Culver, S. J., D. V. Ames, D. R. Corbett, et al. 2006. Foraminiferal and sedimentary record of Late Holocene barrier island evolution, Pea Island, North Carolina: the role of storm overwash. Journal of Coastal Research 22:836–846.

Dolan, R., and H. Lins. 2000. The Outer Banks of North Carolina. U.S. Geological Survey Professional Paper 1177-B.

Dugan, J. E., D. M. Hubbard, M. McCrary, and M. Pierson. 2003. The response of macrofauna communities and shorebirds to macrophyte wrack subsidies on exposed beaches in southern California. Estuarine, Coastal and Shelf Science 58:133–148.

Dugan, J. E., D. M. Hubbard, H. M. Page, and J. Schimel. 2011. Marine macrophyte wrack inputs and dissolved nutrients in beach sands. Estuary and Coasts 34:839–850.

Ellers, O. 1995. Behavioral control of swash-riding in the clam *Donax variabilis*. Biological Bulletin 189:120–127.

Godfrey, P. J., and M. M. Godfrey. 1979. Barrier island ecology of Cape Lookout National Seashore and vicinity. National Park Service Scientific Monograph Series 9, Washington, D.C.

Hancock, T. E., and P. E. Hosier. 2003. The ecology of the endangered species *Amaranthus pumilus*. Castanea 68:236–244.

Hosier, P. E. 2018. Seacoast Plants of the Carolinas. University of North Carolina Press, Chapel Hill. [Now the standard work for coastal botany.]

Hoyt, J. H. 1967. Barrier island formation. Geological Society of America Bulletin 78:1125–1136.

Leatherman, S. P. 1988. Barrier Island Handbook. Laboratory for Coastal Research, University of Maryland, College Park.

Leidner, A. K., and N. M. Haddad. 2010. Natural, not urban, barriers limit dispersal of a coastal endemic butterfly. Conservation Genetics 11:2311–2320.

Oosting, H. J., and W. D. Billings. 1942. Factors affecting vegetational zonation on coastal dunes. Ecology 23:131–142.

Pearse, A. S., H. J. Humm, and G. W. Wharton. 1942. Ecology of sand beaches at Beaufort, North Carolina. Ecological Monographs 12:135–190.

Pilkey, O. H., W. J. Neal, S. R. Briggs, et al. 1998. The North Carolina Shore and Its Barrier Islands. Duke University Press, Durham, N.C.

Shadow, R. A. 2007. Plant Fact Sheet for Sea Oats (*Uniola paniculata* L.). USDA Natural Resources Conservation Service, East Texas Plant Material Center, Nacogdoches.

Susman, V. R., and S. D. Heron Jr. 1979. Evolution of a barrier island, Shackleford Banks, Carteret County, North Carolina. Geological Society of America Bulletin 90:205–215.

Wagner, R. H. 1964. The ecology of *Uniola paniculata* L. in the dune-strand habitat of North Carolina. Ecological Monographs 34:79–96.

Shaped by Salt: Maritime Forests

Au, S.-F. 1974. Vegetation and ecological processes on Shackleford Bank, North Carolina. National Park Service Scientific Monograph Series No. 6. [Includes "ghost trees."]

Bellis, V. J. 1995. Ecology of maritime forests of the southern Atlantic coast: a community profile. Biological Report 30, National Biological Survey, Washington, D.C.

Bourdeau, P. F., and H. J. Oosting. 1959. The maritime live oak community in North Carolina. Ecology 40:148–152.

Boyce, S. G. 1954. The salt spray community. Ecological Monographs 24:29–67.

Sherrill, B. L., A. G. Snider, and C. S. DePerno. 2010. White-tailed deer on a barrier-island: implications for preserving an ecologically important maritime forest. Proceedings Annual Conference of the Southeastern Association of Fish and Wildlife Agencies 64:38–43.

Wells, B. W. 1939. A new forest climax: the salt spray climax of Smith Island, N.C. Bulletin of the Torrey Botanical Club 66:629–634. [Today's Bald Head Island.]

Wells, B. W., and I. V. D. Shunk. 1938. Salt spray: an important factor in coastal ecology. Bulletin of the Torrey Botanical Club 65:485–492. [The first report that salt aerosols, not wind, shape the canopy of maritime forests.]

Wood, V. S. 1981. Live Oaking: Southern Timber for Tall Ships. Northeastern University Press, Boston.

A Rocky Spot on the Beach

Carson, R. 1955. The Edge of the Sea. Houghton Mifflin, Boston. [A literary treasure.]

Frankenberg, D. 2012. The Nature of North Carolina's Southern Coast. 2nd ed. University of North Carolina Press, Chapel Hill.

Paine, R. T. 1966. Food web complexity and species diversity. American Naturalist 100:65–75. [A landmark in the field of ecology that founded the concept of keystone species and trophic cascades.]

Wells, B. W. 1944. Origin and development of the lower Cape Fear Peninsula. Journal of the Elisha Mitchell Scientific Society 60:129–134.

Wilson, A. F. 1995. A history of the Sedgeley Abbey site in New Hanover County. Research Report, North Carolina Department of Cultural Resources, Raleigh.

Jockey's Ridge: A Dune to Remember

Cheplick, G. P., and K. Grandstaff. 1997. Effects of sand burial on purple sand-grass (*Triplasis purpurea*): the significance of seed heteromorphism. Plant Ecology 133:79–89.

Judge, E. K., M. G. Courtney, and M. F. Overton. 2000. Topographic analysis of dune volume and position, Jockey's Ridge State Park, North Carolina. Shore and Beach 68:19–24.

Mitasova, H., M. Overton, and R. S. Harmon. 2005. Geospatial analysis of a coastal sand dune field evolution: Jockey's Ridge, North Carolina. Geomorphology 72:204–221.

Perry, S. 2013. Where sand meets sky: Jockey's Ridge. Our State 80(8):122–130.

Runyon, K. B., and R. Dolan. 2001. Origin of Jockey's Ridge, North Carolina: the end of the highest dune on the Atlantic Coast. Shore and Beach 69:29–32.

Schafale, M. 2012. Live dune barrens. *In* Guide to the Classification of the Natural Communities of North Carolina (4th approximation). North Carolina Natural Heritage Program, Department of Natural and Cultural Resources, Raleigh.

Sorrie, B. A. 2014. Jockey's Ridge State Park. Pages 121–122 *in* An Inventory of the Natural Areas of Dare County, North Carolina. North Carolina Natural Heritage Program, Department of Environment and Natural Resources, Raleigh.

Forces of Nature: Hurricanes and Nor'easters

Barnes, J. 2010. Hurricane Hazel in the Carolinas. Acadia, Charleston, S.C.

Barnes, J. 2013. North Carolina's Hurricane History. 4th ed. University of North Carolina Press, Chapel Hill.

Chaplin, J. E. 2006. The first scientific American, Benjamin Franklin and the pursuit of genius. Basic Books, New York. [See page 122 regarding Franklin's observations of a nor'easter's winds and movement.]

Davis, R. E., and R. Dolan. 1993. Nor'easters. American Scientist 81:428–439. [A primary source of information.]

Dolan, R., and R. E. Davis. 1992. Rating northeasters. Mariners Weather Log 36(1):4–11.

Greening, H., P. Doering, and C. Corbett. 2006. Hurricane impacts on coastal ecosystems. Estuaries and Coasts 29:877–879.

Hudgins, J. E. 2000. Tropical cyclones affecting North Carolina since 1586—an historical perspective. Technical Memorandum NWS ER-92, National

Oceanic and Atmospheric Administration, Washington, D.C. [Includes observations of conditions and damage.]

Knutson, T., S. J. Camargo, J. C. L. Chan, et al. 2019. Tropical cyclones and climate change assessment: part II. Projected response to anthropomorphic warming. https://doi.org/10.1175/BAMS-D-18-0194.1 [Indicates increasing rainfall and greater intensity of future hurricanes, more of which will be category 4 and 5 storms.]

Lohr, S. M., W. E. Taylor, and J. C. Watson. 2004. Restoration, status, and future of the red-cockaded woodpecker on the Francis Marion National Forest thirteen years after Hurricane Hugo. Pages 230–237 *in* Proceedings of the Fourth Red-Cockaded Woodpecker Symposium (R. Costa and S. J. Daniels, eds.). Hancock House, Savannah, Ga.

Stick, D. 1987. The Ash Wednesday Storm. Gresham, Glasgow, U.K.

Beacons, Birds, and Bureaus

Bolen, E. G. 2010. Beacons and birds. Wildlife in North Carolina 74(4):10–15.

Merriam, C. H. 1877. A review of the birds of Connecticut, with remarks on the habits. Transactions of the Connecticut Academy of Arts and Sciences 4:1–165.

Sterling, K. B. 1977. Last of the Naturalists: The Career of C. Hart Merriam. Arno Press, New York.

Mustangs on a Sandy Shore

Blythe, W. B. 1983. The banker ponies of North Carolina and the Ghyben-Herzberg principle. Transactions of the American Clinical and Climatological Association 94:63–72.

Conant, E. K., R. Juras, and E. G. Cothran. 2012. A microsatellite analysis of five colonial Spanish horse populations of the southeastern United States. Animal Genetics 43:53–62.

Gruenberg, B. U. 2015. The Wild Horse Dilemma: Conflicts and Controversies of the Atlantic Coast Herds. Quagga Press, Strasburg, Pa.

Prioli, C. A. 2007. Wild Horses of Shackleford Banks. John F. Blair, Winston-Salem, N.C.

Wood, G. E., M. T. Mengak, and M. Murphy. 1987. Ecological importance of feral ungulates at Shackleford Bank, North Carolina. American Midland Naturalist 118:236–244. [Includes food habits and the impacts of grazing pressure.]

Sea Turtles: Ancient Reptiles on a Modern Shore

Lohmann, K. J., B. E. Witherington, C. M. F. Lohmann, and M. Salmon. 1997. Orientation, navigation, and natal beach homing in sea turtles. Pages 107–135 *in* The Biology of Sea Turtles, Vol. 1 (P. L. Lutz and J. A. Musick, eds.). CRC Press, Boca Raton, Fla.

Lorne, J., and M. Salmon. 2007. Effects of exposure to artificial lighting on orientation of hatching sea turtles on the beach and in the ocean. Endangered Species Research 3:23–30.

McClellan, C. M., and A. J. Read. 2007. Complexity and variation in loggerhead sea turtle life history. Biology Letters 3:592–594.

Miller, J. D., C. J. Limpus, and M. H. Godfrey. 2003. Nest site selection, oviposition, eggs, development, hatching, and emergence of loggerhead turtles. Pages 125–143 *in* Loggerhead Sea Turtles (A. B. Bolten and B. E. Witherington, eds.). Smithsonian Books, Washington, D.C.

Reneker, J. L., and S. J. Kamel. 2016. Climate change increases production of female hatchlings at a northern sea turtle rookery. Ecology 97:3257–3264.

Webster, W. D., and K. A. Cook. 2001. Intraseasonal nesting activity of loggerhead sea turtles (*Caretta caretta*) in southeastern North Carolina. American Midland Naturalist 145:66–73.

Snow Geese on the Outer Banks

Ankney, C. D. 1996. An embarrassment of riches: too many geese. Journal of Wildlife Management 60:217–227. [A call for action to employ previously unacceptable methods in order to regulate waterfowl populations, including greatly increased or no daily bag limits.]

Bolen, E. G., and M. K. Rylander. 1978. Feeding adaptations in the lesser snow goose (*Anser caerulescens*). Southwestern Naturalist 23:158–161.

Giroux, J.-F., G. Gauthier, G. Costanzo, and A. Reed. 1998. Impact of geese on natural habitats. Pages 32–57 *in* The Greater Snow Goose: Report of the Arctic Goose Habitat Working Group (B. D. J. Batt, ed.). Arctic Goose Joint Venture Special Publication. U.S. Fish and Wildlife Service, Washington, D.C., and Canadian Wildlife Service, Ottawa.

Glazener, W. C. 1946. Food habits of wild geese on the Gulf Coast of Texas. Journal of Wildlife Management 10:322–329. [Includes a description of "grubbing."]

Lefebvre, J., G. Gauthier, J.-F. Giroux, et al. 2017. The greater snow goose *Anser caerulescens atlanticus*: managing an overabundant population. Ambio 46 (Suppl. 2):262–274.

Smith, T. J., III, and W. E. Odum. 1981. The effects of grazing by snow geese on coastal salt marshes. Ecology 62:98–106.

Tsang, A., and M. A. Maun. 1999. Mycorrhizal fungi increase salt tolerance of *Strophostyles helvola* in coastal foredunes. Plant Ecology 144:159–166.

Pheasants in the Dunes

Dutton, C. S., and E. G. Bolen. 2000. Fall diet of a relict pheasant population in North Carolina. Journal of the Elisha Mitchell Scientific Society 116:41–48.

Ferris, A. L., E. D. Klonglan, and R. C. Nomsen. 1977. The ring-necked pheasant in Iowa. Iowa Conservation Commission, Des Moines. [Includes pheasant diets in the Midwest.]

Fussell, J. O., III. 1994. A birder's guide to coastal North Carolina. University of North Carolina Press, Chapel Hill.

Marsh, M. 2005. Diminishing returns. Wildlife in North Carolina 69(12):4–8. [Includes information on the rigors of hunting pheasants on the Outer Banks.]

Place, A. R., and E. W. Stiles. 1992. Living off the wax of the land: bayberries and yellow-rumped warblers. Auk 109:334–345.

Quay, T. L. 1959. Unpublished field notes for Hatteras Island. Provided by J. F. Parnell. [Describes the diggings of pheasants in search of pennywort rhizomes.]

Wilson, J. 2005. Bird of beauty, bird of mystery. Wildlife in North Carolina 69(12):9.

Infobox 2-1: Armor-Clad Shorelines

Defeo, O., A. McLachlan, D. S. Schoeman, et al. 2009. Threats to sandy beach ecosystems: a review. Estuarine, Coastal and Shelf Science 81:1–12.

Dugan, J. E., and D. M. Hubbard. 2006. Ecological responses to coastal armoring on exposed sandy beaches. Shore and Beach 74:10–16.

Dugan, J. E., D. M. Hubbard, I. Rodil, et al. 2008. Ecological effects of coastal armoring on sandy beaches. Marine Ecology 29:160–170. [Discusses effects on foraging shorebirds.]

Gittman, R. K., F. J. Fodrie, A. M. Popowich, et al. 2015. Engineering away our natural defenses: an analysis of shoreline hardening in the U.S. Frontiers in Ecology and the Environment 13:301–307.

Gonzalez, A., A. Lambert, and A. Ricciardi. 2008. When does ecosystem engineering cause invasion and species replacement? Oikos 117:1247–1257.

Michaud, K. M., K. A. Emery, J. E. Dugan, et al. 2019. Wrack resource use by intertidal consumers on sandy beaches. Estuarine, Coastal and Shelf Science 221:66–71.

Pilkey, O. H., and J. A. G. Cooper. 2014. The Last Beach. Duke University Press, Durham, N.C.

Riggs, S. R., D. V. Ames, S. J. Culver, and D. J. Mallinson. 2011. The battle for North Carolina's coast. University of North Carolina Press, Chapel Hill.

Young, R., O. Pilkey, D. Heron, et al. 2009. The negative impact of groins. Coastal Care, February 12, 2009. [Twenty-six prominent coastal geologists reject efforts to build "experimental" groins and bypass the law banning hardening structures on North Carolina's coastline.]

Infobox 2-2: Whales and Whaling in North Carolina

Brimley, H. H. 1894. Whale fishing in North Carolina. Bulletin North Carolina Department of Agriculture 14(7):4–8.

Byrd, B. L., A. A. Hohn, G. N. Lovewell, et al. 2014. Strandings as indicators of marine mammal biodiversity and human interactions off the coast of North Carolina. Fishery Bulletin 112:1–23. [Includes data regarding collisions and entanglements with fishing gear.]

Campbell-Malone, R., S. G. Barco, P.-Y. Daoust, et al. 2008. Gross and histologic evidence of sharp and blunt trauma in North Atlantic right whales (*Eubalaena glacialis*) killed by ships. Journal of Zoo and Wildlife Medicine 39:37–55.

Cassoff, R., W. A. McLellan, S. G. Barco, et al. 2011. Lethal entanglement in baleen whales. Diseases of Aquatic Organisms 96:175–185.

Fujiwara, M., and H. Caswell. 2001. Demography of the endangered North Atlantic right whale. Nature 414:537–541.

Johnson, A., G. Salvador, J. Kenney, et al. 2005. Fishing gear involved in entanglements of right whales and humpback whales. Marine Mammal Science 21:635–645.

Kraus, S. D., R. D. Kenney, C. A. Mayo, et al. 2016. Recent scientific publications cast doubt on North Atlantic right whale future. Frontiers in Marine Science 3, article 137. doi:10.3389/fmars.2016.00137.

McElroy, J. 2009. March 1916: the end of North Carolina whaling. This month in North Carolina history, North Carolina Miscellany. Online.

Reeves, R. R., and E. Mitchell. 1988. History of whaling in and near North Carolina. NOAA Technical Report NMFS 65, U.S. Department of Commerce, Washington, D.C.

Simpson, M. B., and S. W. Simpson. 1988. The pursuit of leviathan: a history of whaling on the North Carolina coast. North Carolina Historical Review 65:1–15.

Infobox 2-3: Big, Red, and Good to Eat

McEachron, L. W., R. L. Colura, B. W. Bumguardner, and R. Ward. 1995. Survival of stocked red drum in Texas. Bulletin of Marine Science 62:359–368.

Ramcharitar, J., D. P. Gannon, and A. N. Popper. 2006. Bioacoustics of fishes in the family Sciaenidae (croakers and drums). Transactions of the American Fisheries Society 135:1409–1431.

Ross, J. L., T. M. Stevens, and D. S. Vaughan. 1995. Age, growth, mortality, and reproduction of red drums in North Carolina. Transactions of the American Fisheries Society 124:37–54.

Scharf, F. S., and K. K. Schlicht. 2000. Feeding habits of red drum (*Sciaenops ocellatus*) in Galveston Bay, Texas: seasonal diet variation and predator-prey size relationships. Estuaries 23:128–139.

Stewart, C. B., and F. S. Scharf. 2008. Estuarine recruitment, growth, and first-year survival of juvenile red drum in North Carolina. Transactions of the American Fisheries Society 137:1089–1103.

Infobox 2-4: The Lost Legacy of North Carolina's "Cannery Row"

Anderson, H. 2000. Menhaden Chanteys, an African American maritime legacy. Maryland Marine Notes 18(1):1–6.

Frye, J. 1999. The Men All Singing: The Story of Menhaden Fishing. Expanded 2nd ed. Donning, Virginia Beach, Va.

Garrity-Blake, B. J. 1994. The Fish Factory: Work and Meaning for Black and White Fishermen of the American Menhaden Industry. University of Tennessee Press, Knoxville.

Goodwin, S. 2017. Beyond the Crow's Nest: The Story of the Menhaden Fishery of Carteret County, North Carolina. Self-published, Beaufort, N.C. [Available at the North Carolina Maritime Museum in Beaufort.]

Mims, B. 2014. The fish that built Beaufort. Our State 81(12):44–54.

Trilles, J.-P. 2007. *Olencira praegustator* (Crustacea, Isopoda, Cymothoidae) parasitic on *Brevoorta* species (Pices, Clupeidae) from the southern coasts of North America: review and re-description. Marine Biology Research 3:296–311.

As the sound and smell of the salt marsh are its own, so is its feel.—John and Mildred Teal, *Life and Death of the Salt Marsh*

III

Coastal Marshes

Grasslands of the Coastline: Tidal Salt Marshes

Tidal salt marshes form vast grasslands like no others. They develop under conditions where only a few highly adapted plants can flourish; hence, salt marshes lack the floral diversity typically associated with prairies and other grasslands of the American interior. Two constraints are at work, the first of which is the complete inundation of the roots and lower parts of the plants for several hours about twice a day, each followed by periods of equal length when these same sites lie exposed to the rigors of temporary desiccation and, at times, freezing temperatures. Second, salt in both the soil and water presents intolerable osmotic and toxic difficulties for most vascular plants. Together, these harsh environmental conditions eliminate all but a handful of species—salt-tolerant plants known as halophytes—some of which, lacking competitors, dominate these wetlands in large, monotypic communities.

In North Carolina, as elsewhere, tidal salt marshes typically develop along the mainland where barrier beaches shelter the coastline, whereas others border the protected landward edges of the same offshore formations. In either location, the barrier beaches block wave action that would otherwise prevent water-borne sediments from coalescing into a muddy substrate where marsh plants can gain a foothold (i.e., sites

114

suitable for germination and seedling survival). Once established, marsh vegetation remains resilient against wave action, although the protective value of the marshes may be tempered by strong winds and boat wakes and the advance of rising sea levels.

Two plants dominate the salt marshes along the southeastern coastline of North America (Fig. 3-1). One, smooth cordgrass, forms what is known as low marsh, where tides daily inundate the lower parts of the vegetation. The other, black needlerush, represents high marsh, where tidal waters reach less frequently; salt grass also occupies this zone. Smooth cordgrass in fact dominates low marsh along the entire Atlantic coastline, but north of Delaware Bay, black needlerush gives way to marsh hay as the primary component in high marsh communities. Instead of forming expansive meadows, which in New Jersey and New England were once cut for livestock forage, marsh hay more often occurs on dunes in North Carolina.

Smooth cordgrass may reach heights of six feet along the banks of tidal creeks, whereas large areas of the same species may be half that height elsewhere in the marsh. The difference presumably results because of the deposition of nutrient-rich sediments along the creek edges, although some evidence suggests that genetic influences also may be involved. Ribbed mussels play an important role in cordgrass communities, where they attach themselves to the roots with byssal threads, a protein secretion unaffected by water. The mussels, which may reach densities of several hundred or more per square yard, transform their foods—various forms of plankton—into a natural fertilizer that supplies the plants with important nutrients. In fact, experimental evidence shows that smooth cordgrass grown in the absence of ribbed mussels lacks the vigorous growth of those grown with mussels. The relationship—habitat provided by the plant, nutrition from the animal—represents a type of symbiosis known as facultative mutualism, in which both organisms gain benefits but nonetheless can survive separately. Ribbed mussels also provide an ecological service by stabilizing creek banks and other edges of marsh against erosion.

Black needlerush grows in clumps, often in stands that cover large areas; as with smooth cordgrass, these communities exclude virtually

FIGURE 3-1. Smooth cordgrass dominates North Carolina's salt marshes but commonly shares the tidal zone with clumps or sometimes extensive stands of black needlerush.

all other species of vegetation. In places, they develop as a band along the landward edge of the cordgrass communities. Small patches of these plants will also develop at sites in the marsh where the local relief increases by only an inch (e.g., sites that were once beds of ribbed mussels). For the most part, black needlerush develops where the salinity decreases in comparison with locations occupied by smooth cordgrass. The edges separating the two communities are sharply delineated, and, indeed, abrupt zonation remains the distinctive feature of salt marshes everywhere.

The clumps develop from deep fibrous root systems that offer excellent protection again shoreline erosion; hence, when transplanted, the roots offer a useful tool to restore degraded sites. Little light penetrates through the dense vegetation, which, because of slow decomposition, includes considerable amounts of dead foliage; together, the count of living and dead "needles" may approach a thousand per square yard.

Although not readily apparent, the sharply pointed rigid "stems" of black needlerush actually form from tightly rolled leaves, which in places may reach six feet in height. Smaller plants develop at locations where the salinity increases, and this selective force has apparently resulted in genetic adaptations that do not occur in the taller plants (i.e., taller plants die when they are experimentally relocated to sites where the dwarfed plants thrive). Because of its gray-green color, black needlerush stands out in marked contrast with the emerald hues of smooth cordgrass.

The coarseness of black needlerush severely limits herbivory but offers thick cover for rice rats and marsh rabbits, as well as nesting sites for marsh wrens. Fires readily consume the aboveground foliage but seldom harm the underground rhizomes unless the soil is no longer saturated; in such cases, the flames indeed completely kill the entire stand and severely restrict regrowth. Whereas black needlegrass produces abundant seeds, few seedlings result, and the plants instead rely primarily on rhizomes for reproduction.

A few colorful forbs dot the salt marsh (Fig. 3-2). Among these, sea lavender adds a welcome dash of its namesake hue to the wetland landscape; its flower-bearing stems arise from clumps of oval leaves and reach twelve to twenty-four inches in height. Florists value the clusters

FIGURE 3-2. Wildflowers such as sea lavender, sea oxeye, and perennial saltmarsh aster (*top to bottom*) add occasional touches of color to salt marshes.

of flowers, which, when dried, complement floral arrangements; hence, overharvesting may pose a threat in some areas. Whereas these rhizomatous plants rely heavily on asexual reproduction, which somewhat offsets the effects of their commercial exploitation, they take up to nine years to mature. Many plants thus may be harvested before they can produce the seeds necessary to colonize new locations. Sea oxeye grows in dense patches not unlike a crowded flower bed, where their prominent yellow flowers attract butterflies; the patches usually develop along the upper edges of the tidal zone, where they tolerate both inundation and salt spray. Marsh pink, with its attractive star-shaped flowers—five pink petals with centers of white and yellow— often shares the same habitat as sea lavender and sea oxeye. The small but attractive daisy-like flowers of perennial saltmarsh aster generally occur above the high-tide level. When mature, the plant's small white flowers produce fluffy seed heads that resemble those of dandelions. Other forbs, of course, occur in salt marshes, but for the most part, the verdant expanse of these coastal wetlands lacks the colorful palette and diversity typical of interior grasslands.

The landward edges of salt marshes feature thickets of two shrubs that signal the transition from salt marsh to upland vegetation. Groundsel, also known as silverling or sea myrtle, curiously produces flowers of two colors: in season, female plants display masses of white-plumed seeds that seem like the aftermath of an early snowfall, whereas males produce clusters of small yellow flowers. For the remainder of the year, however, groundsel fades into relative obscurity in the mix of other vegetation in this transition zone. Another member of the aster family, marsh elder, joins groundsel in the same habitat. Robust plants of this species reach heights of eight feet, but those nearest the waterline may be stunted. Unlike groundsel, marsh elder leaves have fuzzy surfaces, three conspicuous veins, and toothed margins. Under normal conditions, both of these shrubs remain confined to a relatively narrow band that is limited by competing vegetation on one side and intolerance to tidal flooding on the other. Both, however, readily exploit new habitat by expanding, for example, along the edges of ditches cut into salt marshes.

Unlike their terrestrial counterparts—the plains and prairies of the American interior—tidal marshes essentially escape the effects of herbivory from hoofed mammals, prairie dogs, and legions of herbivorous insects. Grasshoppers are one of the few grazers that regularly forage on smooth cordgrass and black needlerush, and they do so during the summer months, when they consume considerably less than 1 percent of the available biomass. This means that huge amounts of plant materials avoid consumption and instead decompose, thereby providing a steady flow of recyclable nutrients. This energy flows not only into the marsh itself but also into ecosystems elsewhere, including those offshore, although the export of nutrients is not as great as once believed. Nonetheless, the decomposition of vegetation that otherwise contributes little food value becomes the crucial first step in marine food chains along the Atlantic coast.[1] Thus this source of nutrients in part gave rise to the slogan, "no salt marshes, no seafood," a battle cry for protecting coastal environments from development and pollution.

An Invasive Species Attacks

Common reed—formally designated as *Phragmites australis* but widely known simply as "phrags"—has and continues to commandeer both coastal and inland marshes across the breadth of North America (Fig. 3-3). It comes in three subspecies, one native to much of North America, another occurring along the Gulf Coast, and the last—the

---

1   On the Pacific coast, where salt marshes cover much less area, upwellings enriched by detritus from the seabed provide an abundant source of nutrients to coastal and marine communities. Regrettably, marsh hay and smooth cordgrass have been introduced in some areas on the Pacific shoreline (e.g., San Francisco Bay), where both soon became troublesome. Smooth cordgrass, in particular, invaded mud flats, where it ruined crucial feeding areas for shorebirds in search of invertebrates.

FIGURE 3-3. An aggressive type of common cane, or "phrags," invades eastern salt marshes, where it often completely replaces other vegetation. As shown here, the spring growth of phrags advances into a patch of black needlerush in North Carolina.

aggressive troublemaker—arrived as an import from Europe. In one form or another, phrags thrives on every continent except Antarctica, which hints at its broad tolerance for environmental conditions. In short, phrags is one tough, adaptable plant, of which the invasive form spreads rapidly, with little to halt its aggressive takeover of native wetland communities.

The European subspecies apparently arrived early in the nineteenth century as seeds hitchhiking in the ballast of ships. It soon gained a foothold, pushing out or hybridizing with the native subspecies. Once established, it begins spreading without constraints; only deep water and full-strength seawater stem its advance. Uniquely, phrags thrives within the full spectrum of plant succession—pioneer to climax—in which it crowds out virtually all other plant life, including key species such as smooth cordgrass, black needlerush, and marsh hay. The result is a monoculture of twelve-foot-tall grasses that spread primarily from the lateral growth of rhizomes, which may advance ten feet in length each year and eventually extend sixty feet from the mother plant. Some evidence suggests that phrags also produces toxic substances that inhibit the growth of other plants. The matter remains unclear, but if true, it represents an example of "allelopathy," the adverse effect on plants of one species by the chemical secretions of another.

Phrags modifies coastal marshes in other ways, including stemming the flow of small tidal creeks. When its rhizomes cross these arteries, they entrap sediments, forming dams that effectively curtail further movement of water into the marsh. In a similar fashion, the dense stands of phrags elevate the marsh floor, likewise altering the hydrology of the wetland and, in turn, impacting the fauna dependent on regular tidal inundation. Both of these developments curtail the movements of small fishes and markedly alter the arthropod fauna. For example, entire trophic levels of herbivores and their predators (e.g., spiders) diminish, largely replaced by detritivores. Most of the displaced arthropods represent free-living insects readily incorporated into the food web (e.g., prey for birds), whereas single and much less accessible stem-feeding insects dominate the fauna associated with phrags.

Fires temporarily remove the aboveground growth of phrags, after which the underground parts quickly sprout and renew the stand to its previous state. Some herbicides, if applied repeatedly, will kill the plants, but the expense of spraying the large areas already covered with phrags precludes the practicality of such treatments.

Home for Invertebrates

Salt marshes provide habitat for numerous invertebrates, several of which occur in large numbers, but few are more abundant—and eye-catching— than fiddler crabs and two species of snails (Fig. 3-4). Aptly named, mud fiddler crabs scurry about the marsh floor during high tide in search of food and then recede into their burrows when the water returns, plugging the passageways behind them. Their unbranched burrows may extend two feet deep into the muddy substrate. In winter, fiddlers remain in their burrows, where they enter into a hibernation-like state until warm weather returns. With their dime-sized openings, the burrows aerate the soggy soils at low tide; hence, the improved oxygenation contributes to the well-being of smooth cordgrass. The root systems of smooth cordgrass in turn provide structural support for the burrows in a manner not unlike timbers in a mine shaft, enabling the crabs to excavate at sites where their burrows might otherwise collapse.

Mud fiddler crabs feed on algae using their first (of five) pair of legs, which are outfitted with spoon-like pincers; females use both of these pincers to gather in food. Males, however, make do with just one, as the other claw (either right or left) has evolved into the large "fiddle" used for courtship and territorial defense—and size matters in both situations. Based on studies of a closely related species on the Gulf Coast, the crabs' enlarged claws also dissipate heat, which allows the males to spend more time to feed and display above ground before they retreat to their burrows.

At high tide, marsh periwinkles abandon the exposed muddy areas on the marsh floor and climb the stems of smooth cordgrass. Biologists once believed that the snails fed on algae and detritus on the marsh floor and ascended the grasses solely to avoid crabs and other predators that arrived with the incoming tide. However, recent studies have revealed that marsh periwinkles do in fact graze on the cordgrass, but they also become "fungus farmers." Their husbandry begins when they cut into the leaves of cordgrass with their radulae, which are tongue-shaped structures with a rasp-like surface. In practice, however, grazing becomes less important for gleaning food from the plants' tissues and more important for creating a wound for fungal growth—a farm the periwinkles fertilize with their nitrogen-rich feces. Concurrently, hyphae in the snails' feces help disperse the fungi to other locations in the marsh. The relationship—food for the periwinkles, new habitat for the fungus—offers an example of facultative mutualism, a beneficial but not mandatory relationship between two organisms. Nonetheless, field experiments have shown that the interacting effects of grazing and fungal infections can suppress the growth of smooth cordgrass.

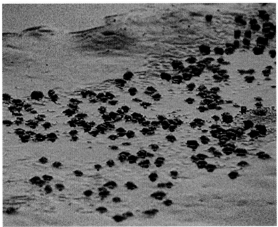

FIGURE 3-4. Invertebrates such as fiddler crabs, mud snails, and marsh periwinkles (*top to bottom*) occur in large numbers in salt marshes.

Large numbers, sometimes in clusters of more than a thousand per square yard, of eastern mud snails thrive on mud flats within salt marshes. No more than an inch in length, the dark shells of these snails develop in spirals of six to seven whorls with a rough surface covered with a slimy veil of algae and mud. They feed primarily on the microflora, especially favoring algae known as diatoms, living in the sediments but also scavenging on the carcasses of other invertebrates or fish. In fact, experimental evidence suggests that mud snails actually may require some degree of animal matter in their diet and thus should be considered as obligate omnivores. On the approach of winter, mud snails migrate to deeper water and enter a period of quiescence, often in aggregations of many thousands. The length of this inactive period likely varies with the severity of winter temperatures; hence, it may be quite short in North Carolina when compared with areas farther north.

Whereas clapper rails feed on mud snails, they seldom appear in the diet of diamondback terrapins, which otherwise prey on snails of a similar size. As suggested by laboratory experiments, the reason seems to be that the shells of mud snails are two to three times more resistant to crushing than other foods in the terrapins' diet, including marsh periwinkles. Predators commonly avoid potential prey that have evolved chemical or physical defenses (e.g., toxic excretions or spines), but here the energetic cost of breaking open the snails outweighs the benefits of their availability and food value. In short, the payoff is not worth the effort, and the terrapins thus seek other foods. In contrast, clapper rails swallow the snails whole and rely on their muscular gizzards to crush the shells.

A Few Saltmarsh Specialists

Salt marshes provide exclusive year-round habitat for relatively few vertebrates. One of these, the diamondback terrapin, has long attracted the attention of naturalists, beginning with John White, who included the species in the sixteenth-century folio of illustrations that he compiled at Roanoke Island. Strangely, White added a sixth toe to the feet on the animal's right side, but his watercolor painting otherwise clearly represents

FIGURE 3-5.
Diamondback
terrapins, once hunted
as a delicacy, are now
a protected species in
North Carolina.

its distinctive features. These include the prominent pattern of concentric-ringed plates, known as scutes, on their carapace (upper shell) and their broad, webbed feet, which, unlike the appendages of sea turtles, have not evolved into flippers (Fig. 3-5).

Up close, diamondback terrapins can be further identified by black flecking on their legs, necks, and heads; the faces of some also sport a prominent black "moustache." Two subspecies (of seven), the northern and Carolina, occur in North Carolina; the range of the northern subspecies extends from Cape Cod, Massachusetts, south to Cape Hatteras, where it intergrades with the Carolina subspecies, which ranges down the coast to northeastern Florida. However, even experts experience difficulty distinguishing these and the other subspecies from one another, in part because of the considerable variation in the appearance of individual terrapins.

Females are noticeably larger than males (carapace lengths of 6–9 vs. 4–5.5 inches), a relationship known as reverse sexual dimorphism, which shows up in only a few other groups of vertebrates (e.g., hawks and owls). They feed on crabs, marine worms, and snails, especially marsh periwinkles. The terrapins' own tasty meat once led to their heavy exploitation as table fare in pricy restaurants. The terrapins commanded as much as one dollar per inch in the marketplace—big money for pockets in the

Greenheads: The Makings of a Bad Day

Saltmarsh horse flies, better known simply as greenheads for their bulging emerald-green eyes, may quickly end a day at the beach or a birding safari near a salt marsh—the half-inch-long flies are all-stars in the painful sport of insect bites. Females begin the annual cycle when they lay eggs on the stems of smooth cordgrass, the dominant vegetation in coastal salt marshes. After hatching, the

Photo credit: Cape Cod Greenhead Fly Control District

larvae fall to the marsh floor, where they prey on invertebrates, including others of their own kind, and remain for the winter concentrated above the high-water mark. Depending on the availability of food, one or two years may pass before the larvae pupate and emerge as adults with a life span of three to four weeks. Initially, the adults of both sexes feed on nectar, but females rely on the carryover of larval fat to produce their first batch of one hundred to two hundred eggs. That done, however, each additional batch of eggs requires a separate meal of protein-rich blood; based on laboratory studies, 70 percent of the females given access to blood produce two additional clutches per summer. This strategy assures production of at least one generation of offspring in an environment that generally lacks large mammals as sources of blood. Hence, to

continue breeding, females sometimes travel inland in search of large-bodied mammals—some with two feet.

Female greenheads, as might be expected, come equipped with mouth parts significantly different from those of males. Their blade-like mandibles act as scissors that cut out wounds in which blood accumulates, as opposed to the needle-like proboscis employed by female mosquitoes, which pierces the skin of their victims without tissue damage. However, the pain from a greenhead bite results not from the wound itself but from the host's reaction to the saliva the females secrete as an anticoagulant. The latter flows from salivary glands through an open duct that empties near the tip of a tongue-like structure and thence into the wound, whereas in males the saliva duct remains closed. Greenhead numbers peak during the summer months, and with a west wind they can easily clear a beach of scantily clad humans, bug sprayed or not. The good news is that only half of the greenhead population become pests, and, unlike mosquitoes, the biters do not transmit human diseases (e.g., malaria and West Nile virus).

Greenheads and other species of biting flies also harass feral horses on barrier islands, often affecting resting and feeding behavior (including nursing foals) and habitat selection. When attacked, horses within a group tightly cluster side by side, which reduces the exposed surface area of each individual; horses on the edges continuously try to move toward the middle, where both of their sides can be protected. Indeed, the number of bites per horse decreases as group size increases, but there is no evidence that the horses form larger groups when the flies become especially troublesome. They also position themselves so that the tail of one horse whisks flies away from the head of another. Moving horses attract more flies than those standing still, thus affecting grazing when the flies abound. The horses also move to beaches and dunes where stronger winds restrict the flies; the lack of vegetation at these sites also reduces the cover available to the flies.

As a countermeasure to greenheads, entomologists have designed simple but effective traps—boxes with an open bottom, a screen top, and a one-way exit into a container. Each box, painted black and erected on four legs about two feet aboveground, services about a quarter acre of territory. The flies, which mistake the boxes for a four-legged animal—deer or livestock—enter from the bottom and, attracted by the light above, move upward inside the box and through the one-way exit, where they become trapped in the container. Bait is not required—the bloodlust of the female greenheads alone provides enough incentive. Box designs vary somewhat, but all rely on attracting the flies into a box from which they cannot escape. A single trap may capture hundreds and sometimes thousands of greenheads per day, but their removal has no lasting effect on the population, as the females have already laid up to two hundred eggs *before* entering a trap. Instead, the traps cut down on the number of bites, perhaps making coastal life a bit more pleasant.

early 1900s. Such pressure soon diminished their numbers, and most coastal states eventually provided some legal protection, but in North Carolina their current status as a species of special concern does not grant as much security as might be desired. The World Conservation Union, for example, designates diamondback terrapins as threatened.

Threats continue not only from degraded marsh habitat but also because the terrapins often enter and drown in crab traps; the small males more easily fit through the entrances, resulting in a population skewed toward females. Abandoned or lost crab traps—so-called ghost traps—continue drowning terrapins for years. In Georgia, for example, two abandoned traps contained the remains of 133 terrapins, mostly males, based on carapace measurements of the accumulated shells. Beginning in 2021, new equipment designed to exclude the terrapins yet still catch and retain crabs must be attached to traps set at two designated management areas in North Carolina, one at Bald Head Island and the other at Masonboro Island; together they include more than 15,600 acres of terrapin habitat. The excluder devices, which can be added to existing traps, change the orientation of the entrances from horizontal to vertical, which keeps out the terrapins but not the more agile crabs. At some locations along the Atlantic coastline, roadkills of females seeking nesting sites also become severe enough to alter the structure of local populations.

Brackish marshes bordering the inner edges of the Outer Banks and adjacent shores on the mainland provide habitat for the Carolina salt marsh snake, a subspecies of the widely distributed northern water snake. Compared with the latter, individuals in this population have dark, almost black dorsal coloration, highlighted by pale crossbands and black half-moon markings on their undersides. While not venomous (although they are often killed because of their similarity with cottonmouths), these and other water snakes nonetheless readily bite when threatened and discharge foul-smelling musk when handled. Although they primarily feed on fish, water snakes do not threaten fish populations and may actually improve fishing by culling diseased or other abnormal individuals from the population.

Salt marsh specialists also include a few birds, among them the saltmarsh sparrow, a migratory species that overwinters in North Carolina

FIGURE 3-6. Seaside sparrows remain year-round in eastern salt marshes, where they occur in separate populations, some of which face an uncertain future.

and elsewhere along the southern coastline and breeds in the marshes of northern states from New Jersey to New England. Another migratory species, the Nelson's sparrow, breeds well inland (e.g., Manitoba) and likewise overwinters along the southern coast.

In contrast, the nonmigratory seaside sparrow remains in place as a year-round resident (Fig. 3-6). However, the distribution of the species along the coastline is not always continuous. Hence, some isolated populations have evolved into subspecies (perhaps as many as nine), including the now-extinct dusky seaside sparrow and the critically endangered Cape Sable seaside sparrow. Isolation heightens the risk of extinction not only for populations on islands but also for those representing "genetic islands" in a "sea" of vulnerable habitat.

In optimal situations, seaside sparrows forage in areas with exposed mud and sparse vegetation near their nesting sites, but if situations are not optimal, then the birds will commute for considerable distances in search of food. Feeding areas also include wrack lines and creek edges. Stands of black needlerush, although prevalent, generally lack desirable foods, and the birds instead forage in smooth cordgrass. Seaside sparrows feed on seeds, spiders, and insects, often beetles. They also probe for small crustaceans, marine worms, and other mud-dwelling invertebrates, a practice in keeping with their longer-than-average (for a sparrow) bills.

Seaside sparrows nest just above the usual extent reached by a spring tide, so named not for the season but because the water "springs" to high levels twice each month on the occasion of a full or new moon. The females alone build and attach their cup-like nests to the base of high-marsh vegetation, commonly black needlerush, often just a few inches aboveground, but other nests may be placed under tidal debris. Most nests are covered by a canopy constructed from surrounding plants. Nesting seems timed to fall between two consecutive spring tides, so that the full cycle of laying, incubation, hatching, and fledging can be completed before the next spring tide arrives. Still, sudden flooding remains a significant cause of nest failure—up to 50 percent in extreme cases—which the birds compensate for by starting anew with a second or even third nest. In some places, rice rats prey on eggs and nestlings to the point of "pushing" the birds from black needlerush (where rice rats also nest) into less desirable habitat where flooding and other predators become greater threats—a sort of "out of the pan and into the fire" relationship.

Virtually any changes befalling a salt marsh jeopardize these sparrows, whose populations have indeed dropped rapidly coincident with coastal developments. The encroachment of phragmites also claims prime habitat—few birds and indeed most other marsh creatures simply find little food or usable cover in the choking growth of this invasive grass. Moreover, the growing prospect of rising sea levels adds further concern for the future of saltmarsh sparrows. In the short term, however, management with prescribed fires stimulates marsh vegetation in ways that favor better nesting habitat for seaside sparrows.

And a Game Bird, Too

Six species of rails seasonally or occasionally visit North Carolina coastal marshes, but clapper rails, also known as marsh hens, remain year-round; one of these, the king rail, more often occurs in freshwater marshes. Rails, because of their laterally compressed bodies, easily move through the forest of grass stems to forage or to escape from predators; weak fliers, they seldom flush to avoid danger (Fig. 3-7). Their thin bodies also gave rise to the expression "skinny as a rail."

Male clapper rails construct nesting platforms, sometimes topped with grassy domes and outfitted with access ramps, in stands of smooth cordgrass at sites near tidal creeks or ponds. However, those birds starting to nest in early April prefer the dense cover of black needlegrass, likely because cordgrass has not yet regrown to its full stature. Thereafter, virtually all nests occur in cordgrass. In North Carolina, the nesting season peaks between May 22 and June 1 and is progressively later in more northerly marshes (e.g., Virginia and New Jersey). Typical

FIGURE 3-7. A few hunters seek the elusive clapper rail, a saltmarsh bird more often heard than seen, but the scant harvest barely reduces their numbers.

clutches consist of ten eggs, incubated in turn by both parents. Should a nesting effort fail—most often from unusually high tides—a pair will begin again; some renest as many as five times in one season. With favorable conditions, however, a pair may produce two broods per year. In all, nesting success generally exceeds 80 percent. Hurricanes often kill large numbers of clapper rails, but because of their high reproductive rate, the decimated populations soon recover.

Both parents care for their brood, which may be reared in two groups, each tended separately by one of the adults. In some cases, one parent tends a brood from an earlier nest while the other incubates a second clutch. Patches of floating drift serve as brooding sites, but the parents also may build brood nests from the dead stems of cordgrass that float in keeping with the changing tides. When encountering high water, the chicks ride on the back of their parents.

Clapper rails, young and adults alike, eat a variety of foods that they generally secure when low tides expose the marsh floor. Crustaceans, primarily fiddler crabs, top the list, followed by snails, mollusks, insects, and occasionally a small fish. When feeding on fiddler crabs, the birds first nip off the large claw of their male victim—a behavior no doubt learned from a bad experience—whereas they swallow a female "as is." Because these foods have hardened exteriors, the birds regurgitate pellets consisting of exoskeletons and shells, which accumulate at sites where the birds hide after feeding. In winter, seeds become more prevalent in the diet of clapper rails, supplemented by polychaetes living in the muddy marsh floor; these marine worms partially replace crustaceans and other prey that become inactive in cold weather. In September and early October migrants that nest farther north join the resident population of clapper rails in North Carolina.

North Carolina is among the coastal states that list clapper rails as game birds, often with generous seasons and bag limits (e.g., nearly ninety days and fifteen birds per day). However, only a small cadre of dedicated hunters make the effort; hence, the annual harvest represents only an insignificant percentage of the rail population. Hunters wait until a high tide concentrates the birds on the higher parts of the marsh, but even then the per person harvest seldom exceeds five per day. In the past, market hunters along the Atlantic coast felled thousands of rails for fancy restaurants in urban centers; in 1918 the slaughter ended with passage of the Migratory Bird Treaty Act and the initiation of regulated harvests by sport hunters.

Saltmarsh Mammals

Few mammals call salt marshes home. Muskrats thrive in brackish marshes at sites where bulrushes and cattails add some variety to the emergent vegetation, but they are otherwise absent where salinities prevail in greater strength. Raccoons and sometimes mink forage in salt marshes at low tide, but neither can be regarded as full-time residents. However, marsh rice rats and marsh rabbits can lay claim to salt marshes as their principal habitat (Fig. 3-8), although nutria, a troublesome exotic species, joins their ranks in some locations. Even so, both marsh rice rats and marsh rabbits at times venture into adjacent uplands or dunes and in some places occur year-round in freshwater wetlands (e.g., the Great Dismal Swamp).

Marsh rice rats lack flattened tails or other obvious adaptations for aquatic life; only the rather small webs between their toes provide a hint of their considerable swimming abilities, including stints underwater when fleeing predators. Unlike most rodents, which largely subsist on seeds, rice rats prey on snails, insects, fiddler crabs, and other arthropods. At times, rice rats gnaw into the hollow stalks of cordgrass to locate the larvae of stem-boring moths. They also eat bird eggs and nestlings of seaside sparrows, marsh wrens, and Foster's terns, at times seriously decreasing the nesting success of these species. Seeds and other plant materials of smooth cordgrass and pickleweeds form a lesser part of their diet. In the past, owners of rice plantations along the lower Cape Fear River regarded the species as pests, likely because of disturbances rather than because they eat rice seeds. Rice rats construct feeding platforms from grass stems, no doubt at one time including rice, and perch on these when tides flood the marsh floor. The platforms approximate the size of dinner plates, and snail shells sometimes accumulate below the platforms in small piles. Because of their abundance, rice rats

FIGURE 3-8. Well named, marsh rice rats (*left*) and marsh rabbits (*right*) are among the few mammals associated with salt marshes.

themselves become prey for northern harriers, short-eared owls, barn owls, and likely other predators that roam in salt marshes.

When breeding, rice rats construct globular nests about the size of a large grapefruit. The nests, made from shredded cordgrass foliage, are attached to cordgrass stems above the high-water mark and have a single side entrance. Most nests occur in the taller stands of cordgrass growing along the edges of tidal creeks or levees. At times, however, rice rats will occupy the nests of marsh wrens; these can be identified because the wrens weave their nests with intact, not shredded, leaves. However, because the wrens also build several false nests, rice rats do not necessarily hijack an occupied nest. Oddly, rice rats exhibit an unusual susceptibility to periodontitis, which makes them useful subjects for studying the disease in humans.

True to their name, marsh rabbits favor coastal marshes but also occur in wetlands elsewhere. Their dark pelage has a reddish tinge, and their dark tail likewise distinguishes them from cottontail rabbits, as does an irregular area of black hair between their ears. Marsh rabbits have short, broad ears and walk rather than hop when foraging. Because of their unusually long toenails, they leave distinctive tracks on muddy surfaces. As might be expected, marsh rabbits swim well and can float almost entirely submerged, leaving only their eyes and nose above water. Taxonomists recognize three subspecies of marsh rabbits, one of which is listed as endangered because of its isolated distribution in the lower Florida Keys—a "bunny" named for the founder of *Playboy* magazine, Hugh Hefner. Hurricanes pose a significant threat to the latter subspecies, as suggested by counts of fecal pellets on study plots before and after Hurricane Irma; after the storm, the pellets had diminished by as much as 94 percent, likely because of the combined effects of direct mortality and the loss of critical habitat.

Marsh rabbits construct covered nests of marsh grasses and fur in which females produce two to four offspring several times per year. Although blind and helpless at birth, the young begin life well furred. Marsh rabbits also make runways in dense vegetation that lead to depressions ("forms"), in which they rest during the day. Northern harriers and great-horned owls prey on marsh rabbits, as did red wolves in the past (but now also where the wolves have been reintroduced; see Chapter 5).

Marsh rabbits eat a wide variety of foods from both marsh and upland vegetation. Southern seaside goldenrod, pennywort, and marsh hay are year-round staples, but the importance of each shifts seasonally. Other foods include beach bean, camphorweed, and common reed, but their complete diet is far more extensive. Marsh plants occur more commonly in the spring, when they make up about one-third of the foods. Few shrubs occur in the diet, even in winter.

Tidal Creeks and Mummichogs

Tidal creeks meander through salt marshes, where they provide crucial links with the open water of bays and estuaries or the ocean itself (Fig. 3-9). At high tide, a marsh seemingly lacks a drainage system, but low tide reveals a dendritic ("tree-like branching") structure of channels and rivulets in which water crosses the landscape in an endless rhythm of ebb and flow. Broadly speaking, tidal creeks come in two sorts, the first being those that carry fresh water downstream from inland watersheds and, in doing so, commonly import various forms of pollution into the marsh ecosystem. Such runoff may include fecal matter, fertilizers, and industrial wastes. This pollution may be relatively subtle, as when the nitrogen and phosphates in fertilizers accelerate eutrophication—the enrichment of water. At its worst, rapid eutrophication causes sudden algal blooms, leading to oxygen depletion, often followed by fish kills and reduced light penetration in the water column.

Other tidal creeks originate in the upper reaches of the marsh, extending no farther inland than the edge of the bordering uplands. These creeks flush with each tidal cycle and, except for precipitation, receive relatively little freshwater runoff. Their limited watersheds form a dendritic pattern little different in their layout from the drainages of the largest rivers. The smallest of these branches, known as rivulets, drain the marsh floor as the tide recedes (i.e., little outbound water spills over the banks of tidal creeks). Conversely, rising tides may overflow the banks and deposit sediments along the creek bank; these form natural levees several inches in height before tapering off toward the interior. The levees, because of their rich soils, typically nourish taller stands of smooth cordgrass in comparison with the somewhat reduced size of the same species elsewhere in the marsh. (Some biologists believe a genetic influence also may be influencing these differences, although that remains unresolved.)

Tidal creeks provide corridors for the dispersal of aquatic life deep into the marsh for food and cover—movements governed by the daily rhythm of the tides but in part by the season. Notably, the diversity of the fish fauna changes in the upper reaches of tidal creeks in keeping with the seasonal appearances of larval and juvenile forms. Near the mouths of the same creeks, however, the year-round presence of older fish gives rise to populations of greater permanence.

Mummichogs well represent the fish fauna of tidal creeks. Their curious but quaint name originates from a Native American name meaning "going in crowds," a reflection of their tendency to school in large numbers. Mummichogs are one of several species of killifish—as apart from minnows—and tolerate a broad range of salinities, temperature, dissolved oxygen, and pollution. As such, they are valued as "lab rats" for research, including serving as the first species of fish to travel in space.

FIGURE 3-9. Tidal creeks provide mummichogs and other marine life with twice-daily access to food and cover in the interior of salt marshes. Mummichogs, minnow-like but actually a killifish, reach average lengths of 3 to 3.5 inches (*inset*). Note the abrupt but characteristic zonation between smooth cordgrass (bordering the creek) and black needlegrass. Photo credit: Jeffery Merrell (*inset*).

Nutria at Large

Rodents from South America that superficially resemble large muskrats or small beavers carved out a niche in wetlands across the United States following their escape or intentional releases in the 1930s and 1940s. Known as nutria, these twelve-pound rodents produce a coat of thick, rather coarse fur that some thought would become a moneymaker for fur farmers; others naively released the animals as "weed cutters" in hopes of controlling unwanted aquatic vegetation.[1] Neither plan worked out, and those captives that were not intentionally released by unhappy fur farmers escaped when hurricanes destroyed their pens. Nutria thereafter spread rapidly and were first reported in North Carolina in 1958. Today nutria are widely regarded as pests that may require control where they damage wetland vegetation or crops. In some areas, a few trappers still find a market for the bulky pelts.

Unlike the flattened tails of muskrats and beaver, nutria feature long round tails. Webs occur only between the first four digits of their hind feet; the free fifth toe is used for grooming. Nutria walk in a hunched manner, the result of large rear legs and short front legs. Their lips close behind their two large orange-colored incisors to enable gnawing while underwater; likewise, their nostrils seal out water. Mammary glands develop high on the sides of females, which allows the kits to nurse while swimming.

Nutria feed opportunistically on a variety of herbaceous vegetation, primarily wetland plants, and it seems likely that they may compete with muskrats for food where both species occur. They also forage on crops in fields adjacent to marshes and include alfalfa, rice, corn, and sugarcane, among others. At high densities, nutria populations may alter the successional patterns in marshes and destroy desirable wetland vegetation. Such damage occurs because roots and rhizomes form most of their diet; the nutria remove the most vital parts of the plants. Moreover, they consume about 25 percent of their body weight daily. At times, nutria build feeding platforms. With the steady accumulation of vegetation, the structures may rise well above the waterline and thereby resemble muskrat houses.

---

1   To avoid confusion, some prefer "coypu," because in Spanish "nutria" refers to an otter. Nonetheless, "nutria" remains widely used in English-speaking countries.

Where banks with slopes of at least forty-five degrees occur, nutria will dig burrows. Some are simple, but others may be a complex of tunnels and chambers that serve as feedings areas and nurseries. (The tunnels also weaken levees and disrupt drainage systems.) Lacking burrows, they rely on platform nests. They breed year-round, usually producing litters of three to six kits, although sometimes as many as twelve. Litter size diminishes somewhat in winter but increases in areas with mild winters and plentiful food. Under favorable conditions, females produce about three litters per year; the kits arrive fully furred and active. With a reproductive potential of this magnitude, nutria populations quickly mushroomed. In Louisiana alone, for example, the escapees produced a population of twenty million in less than twenty years.

Great horned owls, northern harriers, and red-tailed hawks prey on nutria, as do mink and red foxes. Alligators thrive on nutria in Louisiana; in one area, nutria make up 60 percent by weight of alligators' diet; they likely prey heavily on nutria elsewhere in the Deep South as well. Jaguars represent an important predator of nutria in South America. Cold weather significantly reduces nutria populations, especially at the northern edges of their distribution; in Maryland, for example, a hard winter claimed 90 percent of the population. Elsewhere, including North Carolina, cold snaps can cause severe frostbite, resulting in the loss of tails, toes, and ears. Despite such setbacks, nutria often become a significant ecological factor wherever they occur.

Back down on Earth, however, they have another useful function: consuming considerable numbers of mosquito larvae. Indeed, some estimates suggest that a single fish might feed on as many as two thousand "wigglers" per day. Mummichogs also provide forage for larger fishes of commercial or recreational value, as well as for herons and egrets of several species. Whereas spells of cold weather seldom kill mummichogs, they do impair their responses and thus may increase their vulnerability to fish-eating predators.

Besides mosquito larvae, mummichogs feed on amphipods, copepods, and other small crustaceans, marine worms, fish eggs (including their own), algae, and, at times, detritus, which provides essentially no nutritional benefits. As might be expected, their daily movements back and forth between tidal creeks and the surrounding wetlands coincide with the tides, which maximizes their feeding time in the shallow, food-rich waters covering the marsh at high tide. At spring tide, females produce clutches of several hundred sticky eggs, which they deposit on the underwater surfaces of plants, most commonly those of smooth cordgrass. Females also lay their clutches inside the partially opened, empty shells of ribbed mussels, which offer protection from egg predators; as it happens, the distributions of the mussels and mummichogs fully overlap. When the tide ebbs, the spawning sites will remain exposed for about fifteen days. The eggs thus develop their embryos out of water, which apparently is essential for successful incubation (i.e., as shown experimentally, development halts if the eggs remain underwater). The eggs hatch when the next spring tide again covers the eggs. Females may spawn up to six times per year, each in synchrony with the occurrence of spring tides.

Although mummichogs form an established link in saltmarsh food chains, some unfortunate mummichogs face an even greater risk of predation. An interesting cycle begins when the gills of mummichogs become infected with the larvae of parasitic trematodes newly released from their snail hosts. Presumably to gasp for air, the infected fish rise to the surface, where their jerking movements attract the attention of hungry herons, which easily prey on the distressed fish. In its newest host, the parasite matures and produces eggs, which enter the water in the bird's excreta. Snails then ingest the eggs, which later become the larvae that attach to an unsuspecting mummichog. In all, the trematode's complex life history entails a three-stage cycle of hosts consisting of fish-eating birds, snails, and mummichogs—not bad for a flatworm.

Up the Estuary: Tidal Freshwater Marshes

Tidal freshwater marshes develop at the upper end of estuaries where the influence of salinity barely lingers, if at all, yet water levels still change with the ebb and flow of daily tides. These marshes typically occur where the estuary narrows and magnifies the tidal amplitude. An additional factor prevails in North Carolina: the presence or absence of closely spaced barrier islands that separate estuaries from the ocean. Where present, barrier islands reduce the magnitude of the ocean tides that would otherwise extend into the sounds and continue upstream in the estuaries of rivers such as the Neuse and Chowan. Barrier islands thus hinder the development of significant areas of freshwater tidal marshes on the major rivers in North Carolina. Instead, the edges of estuaries along these rivers develop into freshwater swamps where cypress and tupelo gum dominate the shoreline wetlands in lieu of marsh vegetation. An exception helps prove the rule.

Only the Cape Fear River flows directly into the ocean at a location absent a barrier island between its mouth and the open sea. This means that the ocean's tides extend unimpeded into the river's upper estuary, which is indeed the only place in North Carolina with prominent tidal freshwater marshes. As the tide rises, it pushes back the downstream flow of fresh water, which floods the edges of the estuary on a regular basis. Hence, tides fluctuate four to five feet in the Cape Fear River estuary, whereas the amplitude of lunar tides in the state's other large estuaries rarely exceeds a few inches (although winds at times cause larger tidal amplitudes at these locations). The distribution of tidal freshwater marshes along the Atlantic coastline thus reveals an obvious gap between Virginia and South Carolina—except for the Cape Fear estuary. Regrettably, rising sea levels and dredging have allowed salt water to intrude farther up the Cape Fear River, thereby threatening North Carolina's best example of a tidal freshwater marsh.

In contrast to tidal salt marshes, far more species of plants occur in freshwater tidal marshes. The diversity produces a mix of life forms: submerged plants, floating-leaf plants, broad-leaf emergents, grass-like sedges and rushes, true grasses, and a few shrubs. Examples include watermilfoil, spatterdock, pickerelweed, three-square bulrush, cattails, giant cordgrass, and buttonbush. In all, fifty to sixty species may occur at a single location. At the downstream edges of the marsh where the salinity may slightly increase, the composition of the flora likewise changes somewhat as some species drop out and others appear, but none seems endemic to these marshes (i.e., all occur in freshwater marshes elsewhere). Tidal freshwater marshes lack the distinctive zonation that characterizes tidal salt marshes, although some ecologists recognize subtle differences that represent "high" and "low" zones. For the most part, however, the vegetation appears almost randomly distributed in irregular patterns. The vegetation also undergoes a sequence of seasonal changes: some plants appear in early spring, others appear later, and still others dominate the marshes in late summer.

Tidal freshwater marshes provide habitat for a sizable number of reptiles and amphibians, as well as furbearing mammals, including turtles, snakes, frogs, and muskrats. Likewise, a rich community of insects occurs in these areas, but other invertebrates (e.g., mollusks and crustaceans) are relatively uncommon in comparison with tidal salt marshes. However, the diversity of fishes at these locations lacks the richness of the fish fauna either farther upstream beyond the influence of tidal flow or downstream toward the saline waters near the mouth of the estuary. Many species of birds visit tidal freshwater marshes because of the structural diversity of the habitat. Among these, waterfowl, rails, coots, and wading birds frequent the marsh communities, but shrubs and trees along the upland edges attract many arboreal species as well; swallows and flycatchers, for example, prey on the abundant insect populations. Like the other components of the fauna, however, none of these birds has a unique association with any of the various types of habitat available in tidal freshwater marshes.

Historically, tidal freshwater marshes offered attractive sites for rice plantations, of which twenty-eight flourished along the lower Cape Fear

River during the eighteenth and early nineteenth centuries. Of these, Orton Plantation, now being restored to again grow rice, is perhaps the best known. Like the tidal freshwater marshes, the plantations were limited to a zone far enough upriver to avoid salt water but not far enough to lose the effect of tidal flow—a limitation that turned the plantations into highly valuable real estate. At the time, slaves diked the wetlands and harvested the crops, and rather elaborate wooden gates ("trunks") controlled the water flowing from the river into the fields. The end of slavery also ended the economic importance of rice in the region, although Orton Plantation and a handful of others operated into the early 1900s. Remnants of the abandoned dikes still persist at some locations along the Cape Fear River, whereas at others, notably in South Carolina, the fields have been transformed into well-managed waterfowl habitat. The habitat at Waccamaw National Wildlife Refuge on the Coastal Plain of South Carolina, for example, includes sixty-five hundred acres of tidal freshwater marshes that were once commercial rice fields.

READINGS AND REFERENCES

General

Hosier, P. E. 2018. Seacoast Plants of the Carolinas. University of North Carolina Press, Chapel Hill. [A guide and primary resource for the ecology and vegetation of salt marshes.]

Teal, J., and M. Teal. 1969. Life and Death of the Salt Marsh. Little, Brown, Boston.

Grasslands of the Coastline: Tidal Salt Marshes

Adams, D. A. 1963. Factors influencing vascular plant zonation in North Carolina salt marshes. Ecology 44:445–456.

Bertness, M. D. 1984. Ribbed mussels and *Spartina alterniflora* production in a New England salt marsh. Ecology 65:1794–1807.

Bertness, M. D., and E. Grosholz. 1985. The population dynamics of the ribbed mussel, *Geukensia demissa*: the costs and benefits of an aggregated distribution. Oecologia 67:192–204.

Bilkovic, D. M., M. M. Mitchell, R. E. Isdell, et al. 2017. Mutualism between ribbed mussels and cordgrass enhances salt marsh nitrogen removal. Ecosphere 8:e01795.

Calloway, J. C., and M. N. Joselyn. 1992. The introduction and spread of smooth cordgrass (*Spartina alterniflora*) in south San Francisco Bay. Estuaries 15:218–226.

Christian, R. R., W. L. Bryant Jr., and M. M. Brinson. 1990. *Juncus romerianus* production and decomposition along gradients of salinity and hydroperiod. Marine Ecology Progress Series 68:137–145.

Davis, L. V., and I. E. Gray. 1966. Zonal and seasonal distribution of insects in North Carolina salt marshes. Ecological Monographs 36:275–295.

Franz, D. R. 2001. Recruitment, survivorship, and age structure of a New York ribbed mussel population *Geukensia demissa* in relation to shore level—a nine year study. Estuaries 24:319–327.

Haines, E. B. 1977. The origins of detritus in Georgia salt marsh estuaries. Oikos 29:254–260. [See also Haines, E. B. 1979. Interactions between Georgia salt marshes and coastal waters: a changing paradigm. Pages 35–46 *in* Ecological Processes in Coastal and Marine Systems (R. J. Livingston, ed.). Plenum Press, New York.]

Kemp, P. F., S. Y. Newall, and C. Krambeck. 1990. Effects of filter-feeding by the ribbed mussel *Geukensia demissa* on the water-column microbiota of *Spartina alterniflora* saltmarshes. Marine Ecology Progress Series 50:119–131.

Kneib, R. T. 1997. The role of tidal marshes in the ecology of estuarine nekton. Pages 163–220 *in* Oceanography and Marine Biology: An Annual Review, Vol. 35 (A. D. Ansell, R. N. Gibson, and M. Barnes, eds.). UCL Press, Bristol, Pa. ["Nekton" refers to aquatic animals capable of sustained self-propelled movements in a horizontal direction, including fishes and crustaceans, among others.]

Parsons, K. A., and A. de la Cruz. 1980. Energy flow and grazing behavior of conocephaline grasshoppers in a *Juncus roemerianus* marsh. Ecology 61:1045–1050.

Pennings, S. C., M. B. Grant, and M. D. Bertness. 2005. Plant zonation in low-latitude salt marshes: disentangling the roles of flooding, salinity, and competition. Journal of Ecology 93:159–167. [Deals with black needlerush and smooth cordgrass (i.e., low and high marsh zones).]

Smalley, A. E. 1960. Energy flow of a salt marsh grasshopper population. Ecology 41:672–677. [Grasshopper herbivory of smooth cordgrass.]

Stiven, A. E., and S. A. Gardner. 1992. Population processes in the ribbed mussel *Geukensia demissa* (Dillwyn) in a North Carolina salt tidal marsh gradient: spacial pattern, predation, growth, and mortality. Journal of Experimental Marine Biology and Ecology 160:81–102.

Teal, J. 1962. Energy flow in the salt marsh ecosystem of Georgia. Ecology 43:614–624. [A seminal study heralding the productivity of salt marshes and their role as exporters of nutrients to other coastal communities. Later research (Haines 1977) modified the magnitude of these original estimates.]

Wiegart, R. G., and B. J. Freeman. 1990. Tidal salt marshes of the southeastern Atlantic coast: a community profile. Biological Report 85 (7.29), U.S. Fish and Wildlife Service, Washington, D.C.

Woerner, L. S., and C. T. Hackney. 1997. Distribution of *Juncus roemerianus* in North Carolina tidal marshes: the importance of physical and biotic variables. Wetlands 17:284–291.

An Invasive Species Attacks

Able, K. W., and S. M. Hagan. 2003. Impact of common reed, *Phragmites australis*, on essential fish habitat: influence on reproduction, embryological development and larval abundance of mummichog (*Fundulus heteroclitus*). Estuaries 26:4050.

Chambers, R. M., L. A. Meyerson, and K. Saltonstall. 1999. Expansion of *Phragmites australis* into the tidal marshes of North America. Aquatic Botany 64:261–273.

Gratton, C., and R. F. Denno. 2005. Restoration of arthropod assemblages in a *Spartina* salt marsh following removal of the invasive plant *Phragmites australis*. Restoration Ecology 13:358–372.

Roman, C. T., W. A. Niering, and R. S. Warren. 1984. Salt marsh vegetation changes in response to tidal restriction. Environmental Management 8:141–149.

Saltonstall, K. 2002. Cryptic invasion by a non-native genotype of the common reed, *Phragmites australis*, into North America. Proceedings of the National Academy of Sciences of the United States of America 99:2445–2449.

Uddin, M. N., R. W. Robinson, A. Buultjens, et al. 2017. Role of allelopathy of *Phragmites australis* in its invasion processes. Journal of Experimental Marine Biology and Ecology 486:237–244.

Home for Invertebrates

Bertness, M. D. 1985. Fiddler crab regulation of *Spartina alterniflora* production in a New England salt marsh. Ecology 65:1794–1807.

Curtis, L. A., and L. E. Hurd. 1979. On the broad nutritional requirements of the mud snail, *Ilyanassa (Nassarius) obsoleta* (Say), and its polytrophic role in the food web. Journal of Experimental Marine Biology and Ecology 41:289–297.

Darnell, M. Z., and P. Munguia. 2011. Thermoregulation as an alternate function of the sexually dimorphic fiddler crab claw. American Naturalist 178:419–428.

Grimes, B. H., M. T. Huish, J. H. Kerby, and D. Moran. 1989. Species profiles: live histories and environmental requirements of coastal fishes and invertebrates (mid-Atlantic)—Atlantic marsh fiddler. U.S. Fish and Wildlife Service Biological Report 82 (11.114), Washington, D.C., and U.S. Army Corps of Engineers TR EL-82-4, Vicksburg, Miss.

Lin, J. 1991. Predator-prey interactions between blue crabs and ribbed mussels living in clumps. Estuarine, Coastal and Shelf Science 32:61–69.

Scheltema, R. S. 1964. Feeding habits and growth in the mud-snail, *Nassarius obsoletus*. Chesapeake Science 5:161–166.

Silliman, B. R., and S. Y. Newell. 2003. Fungal farming in a snail. Proceedings of the National Academy of Sciences 100:15643–15648.

Tucker, A. D., S. R. Yeomans, and J. W. Gibbons. Shell strength of mud snails (*Ilyanassa obsolete*) may deter foraging by diamondback terrapins (*Malaclemys terrapin*). American Midland Naturalist 138:224–229.

Vaughn, C. C., and F. M. Fisher. 1988. Vertical migration as a refuge from predation in intertidal marsh snails: a field test. Journal of Experimental Marine Biology and Ecology 123:163–176.

A Few Saltmarsh Specialists

Avissar, N. G. 2006. Changes in population structure of diamondback terrapins (*Malaclemys terrapin*) in a previously surveyed creek in New Jersey. Chelonian Conservation and Biology 5:154–159.

Conant, R., and J. D. Lazell. 1953. The Carolina salt marsh snake: a distinct form of *Natrix sipedon*. Breviora 400:1–13. [Genus *Natrix* now renamed *Nerodia*.]

DiQuinzio, D. A., P. W. C. Paton, and E. R. Eddleman. 2002. Nesting ecology of saltmarsh sharp-tailed sparrows in a tidally restricted salt marsh. Wetlands 22:179–185.

Dorcas, M. E., J. D. Willson, and J. W. Gibbons. 2007. Population decline and demographic changes of a diamondback terrapin population over three decades. Biological Conservation 137:334–340.

Grosse, A. M., J. C. Maerz, H. Hepinstall-Cymeerman, and M. E. Dorcus. 2011. Effects of roads and crabbing pressures on diamondback terrapin populations in coastal Georgia. Journal of Wildlife Management 75:762–770.

Grosse, A. M., J. D. van Dijk, K. L. Holcomb, and J. C. Maerz. 2009. Diamondback terrapin mortality in crab traps in a Georgia tidal marsh. Chelonian Conservation and Biology 8:98–100. [Reports 133 drownings, 83 percent of which were males, in two abandoned crab traps.]

Hunter, E. A. 2017. How will sea-level rise affect threats to nesting success of seaside sparrows? Condor 119:459–468.

Kern, R. A., W. G. Shriver, J. L. Bowman, et al. 2012. Seaside sparrow reproductive success in relation to prescribed fire. Journal of Wildlife Management 76:932–939.

Marshall, R. M., and S. E. Reinert. 1990. Breeding ecology of seaside sparrows in a Massachusetts salt marsh. Wilson Bulletin 102:501–513.

Post, W. 1981. The influence of rice rats *Oryzomys palustris* on the habitat use of the seaside sparrow *Ammospiza maritima*. Behavioral Ecology and Sociobiology 9:35–40.

Post, W., and J. S. Greenlaw. 1994. Seaside sparrow (*Ammodramus maritimus*). The Birds of North America, No. 94 (A. Poole and F. Gill, eds.). Academy of Natural Sciences, Philadelphia, Pa., and American Ornithologists' Union, Washington, D.C.

Post, W., J. S. Greenlaw, T. L. Merriam, and L. A. Wood. 1983. Comparative ecology of northern and southern populations of the seaside sparrow. Pages 123–136 *in* Seaside Sparrow, Its Biology and Management (T. L. Quay, J. B. Funderburg Jr., D. S. Lee, et al., eds.). Occasional Paper North Carolina Biological Survey, North Carolina State Museum, Raleigh.

Roosenburg, W. M., and J. P. Green. 2000. Impact of a bycatch reduction device on diamondback terrapin and blue crab capture in crab pots. Ecological Applications 10:882–889.

Shriver, W. G., T. P. Hodgman, J. P. Gibbs, and P. D. Vickery. 2010. Home range sizes and habitat use of Nelson's and saltmarsh sparrows. Wilson Journal of Ornithology 122:340–345.

Szerlag, S., and S. P. McRobert. 2006. Road occurrence and mortality of the northern diamondback terrapin. Applied Herpetology 3:27–37.

And a Game Bird, Too

Adams, D. A., and T. L. Quay. 1958. Ecology of the clapper rail in southeastern North Carolina. Journal of Wildlife Management 22:149–156.

Eddleman, W. R., and C. J. Conway. 1998. Clapper rail (*Rallus longirostris*). The Birds of North America, No. 340 (A. Poole and F. Gill, eds.). The Birds of North America Inc., Philadelphia, Pa.

Heard, R. W. 1982. Observations on the food and food habits of clapper rails (*Rallus longirostris* Boddaert) from tidal marshes along the eastern and gulf coasts of the United States. Gulf Coast Research Reports 7:125–135.

Meanley, B. 1985. The Marsh Hen: A Natural History of the Clapper Rail of the Atlantic Coast Salt Marsh. Tidewater Publishers, Centerville, Md.

Saltmarsh Mammals

Brunjes, J. H., IV, and W. David Webster. Marsh rice rat, *Oryzomys palustris*, predation on Foster's tern, *Sterna forsteri*, eggs in North Carolina. Canadian Field-Naturalist 117:654–655.

Chapman, B. R., and M. K. Trani. 2007. Marsh rabbit: *Sylvilagus palustris*. Pages 247–252 *in* The Land Manager's Guide to Mammals of the South (M. K. Trani, W. M. Ford, and B. R. Chapman, eds.). The Nature Conservancy, Durham, N.C., and U.S. Forest Service, Atlanta, Ga.

Hamilton, W. J., Jr. 1946. Habits of the swamp rice rat, *Oryzomys palustris palustris* (Harlan). American Midland Naturalist 36:730–736.

Kruchek, B. L. 2004. Use of tidal marsh and upland habitats by the marsh rice rat (*Oryzomys palustris*). Journal of Mammalogy 85:569–575.

Leopard, E. P. 1979. Periodontitus. Animal model: periodontitis in the rice rat (*Oryzomys palustris*). American Journal of Pathology 96:643–646.

Markham, K. W., and W. D. Webster. 1993. Diets of marsh rabbits, *Sylvilagus palustris* (Lagomorpha: Leporidae), from coastal islands in southeastern North Carolina. Brimleyana 19:147–154.

Montalvo, A. E., I. D. Parker, A. A. Lund, et al. 2020. Effects of Hurricane Irma on the endangered lower keys marsh rabbit. Southeastern Naturalist 19:759–770.

Post, W. 1981. The influence of rice rats *Oryzomys palustris* on the habitat use of the seaside sparrow *Ammospiza maritima*. Behavioral Ecology and Sociobiology 9:35–40.

Sharp, H. F., Jr. 1967. Food ecology of the rice rat, *Oryzomys palustris* (Harlen), in a Georgia salt marsh. Journal of Mammalogy 48:557–563.

Stiling, P. D., and D. R. Strong. 1983. Weak competition among *Spartina* stem borers, by means of murder. Ecology 64:770–778.

Tidal Creeks and Mummichogs

Able, K. W., and M. Castagna. 1976. Aspects of an undescribed reproductive behavior in *Fundulus heteroclitus* (Pisces: Cyprinodontae) from Virginia. Chesapeake Science 16:282–284. [First report of mummichog eggs laid in empty ribbed mussel shells.]

Abraham, B. J. 1985. Species profiles: life histories and environmental requirements of coastal fishes and invertebrates (mid-Atlantic)—mummichogs and striped killifish. U.S. Fish and Wildlife Service Biological Report 82 (11.40), Washington, D.C., and U.S. Army Corps of Engineers TR EL-82-4, Vicksburg, Miss.

Allen, D. M., S. S. Haertel-Borer, B. J. Milan, et al. 2007. Geomorphological determinants of nekton use of intertidal salt marsh creeks. Marine Ecology Progress Series 329:57–71.

Bretsch, K., and D. M. Allen. 2006. Tidal migrations of nekton in salt marsh intertidal creeks. Estuaries and Coasts 29:474–486.

Hackney, C. T., W. D. Buranck, and O. P. Hackney. 1976. Biological and physical dynamics of a Georgia tidal creek. Chesapeake Science 17:271–280.

Hettler, W. F., Jr. 1989. Nekton use of regularly flooded saltmarsh cordgrass habitat in North Carolina, USA. Marine Ecology Progress Series 56:111–118.

Kneib, R. T. 1986. The role of *Fundulus heteroclitus* in salt marsh trophic dynamics. American Zoologist 26:259–268.

Kneib, R. T., and A. E. Stiven. 1978. Growth, reproduction, and feeding of *Fundulus heteroclitus* (L.) on a North Carolina salt marsh. Journal of Experimental Marine Biology and Ecology 31:121–140.

Mallin, M. A., and A. J. Lewitus. 2004. The importance of tidal creek ecosystems. Journal of Experimental Marine Biology and Ecology 298:145–149.

Meyer, D. L., and M. H. Posey. 2009. Effects of life history strategy on fish distribution and use of estuarine salt marsh and shallow-water flat habitats. Estuaries and Coasts 32:797–812. [Influences of water depth on mummichogs and other fishes in North Carolina.]

Rozas, L. P., C. C. McIvor, and W. E. Odem. 1988. Intertidal rivulets and creekbanks: corridors between tidal creeks and marshes. Marine Ecology Progress Series 47:303–307.

Sanger, D. M., A. Blair, G. DiDonato, et al. 2015. Impacts of coastal develop-
    ment on the ecology of tidal creek ecosystems of the U.S. Southeast including
    consequences to humans. Estuaries and Coasts 38(Supplement 1):49–66.
Santiago Bass, C., and J. S. Weis. 2009. Conspicuous behavior of *Fundulus
    heteroclitus* associated with high digenean metacercariae gill abundances.
    Journal of Fish Biology 74:763–772. [Describes increased predation resulting
    from infections of flatworm larvae.]
Shenker, J. M., and J. M. Dean. 1997. The utilization of an intertidal salt marsh
    creek by larval and juvenile fishes: abundance, diversity, and temporal varia-
    tion. Estuaries 2:154–163.
Von Baumgarten, R. J., R. C. Simmonds, J. F. Boyd, and O. K. Garriott. 1975.
    Effects of prolonged weightlessness on the swimming pattern of fish aboard
    Skylab 3. Aviation, Space, and Environmental Medicine 46:902–906.
    [Mummichogs in space.]

Up the Estuary: Tidal Freshwater Marshes

Foushee, R. 2001. Cape Fear gold. Wildlife in North Carolina 65(1):4–11.
    [Describes rice plantations in southeastern North Carolina.]
Odum, W. E., T. J. Smith III, J. K. Hoover, and C. C. McIvor. 1984. The ecology
    of tidal freshwater marshes of the United States East Coast: a community
    profile. FWS/OBS-83-17, U.S. Fish and Wildlife Service, Washington, D.C.
Whigham, D. F., A. H. Baldwin, and A. Barendregt. 2019. Tidal freshwater
    marshes. Pages 619–640 *in* Coastal Wetlands: An Integrated Approach
    (G. M. E. Perillo, E. Wolanski, D. R. Cahoon, and C. S. Hopkinson, eds.).
    2nd ed. Elsevier, Cambridge, Mass.

Infobox 3-1: Greenheads: The Makings of a Bad Day

Bolsler, E. M., and E. J. Hansens. 1974. Natural feeding behavior of adult salt-
    marsh greenhead, and its relation to oogenesis. Annals of the Entomological
    Society of America 67:321–324.
Magnarelli, L. A., and J. G. Stoffolano Jr. 1980. Blood feeding, oogenesis,
    and oviposition by *Tabanus nigrovittatus* in the laboratory. Annals of the
    Entomological Society of America 73:14–17.
Powell, D. M., D. E. Danze, and M. A. Gwinn. Predictors of biting fly harass-
    ment and its impact on habitat use by feral horses (*Equus caballus*) on a
    barrier island. Journal of Ethology 24:147–153.

Stoffolano, J. G., Jr., and L. R. S. Yin. 1983. Comparative study of the mouth-
parts and associated sensilla of adult male and female *Tabanus nigrovittatus*
(Diptera: Tabanidae). Journal of Medical Entomology 20:11–32.

Wall, W. J., and O. W. Owen Jr. 1980. Large scale use of box traps to study
and control saltmarsh greenhead flies (Diptera: Tabanidae) on Cape Cod,
Massachusetts. Environmental Entomology 9:371–375. [Describes one of
several trap designs.]

Infobox 3-2: Nutria at Large

Brown, L. N. 1975. Ecological relationships and breeding biology of the nutria
(*Myocaster coypus*) in the Tampa, Florida, area. Journal of Mammalogy
56:928–930.

Evans, J. 1970. About nutria and their control. Resource publication 86:1–65,
U.S. Bureau of Sport Fisheries and Wildlife, Washington, D.C.

Willner, G. R., J. A. Chapman, and D. Pursley. 1997. Reproduction, physiological responses, food habits and abundance of nutria on Maryland marshes.
Wildlife Monographs 65:1–43.

Woods, C. A., L. Contreras, G. Willner-Chapman, and H. P. Widden. 1992.
Myocaster coypus. Mammalian Species No. 398. American Society of
Mammalogists.

All rivers feed the sea, true, but in our world they all but one first feed the sounds.—Bland Simpson, *Into the Sound Country*

Rivers, Estuaries, and Sounds

Rivers and sounds figured heavily in the early history of North Carolina. The state's oldest communities, whether successful or not, began on shores where natural resources seemed plentiful and commerce might flourish. These include Bath, New Bern, and Edenton, as well as the star-crossed colony at Roanoke and the ghostly remains of Brunswick Town. Still, for naturalists, the ecology of these waters and their biota prove the equal of pirates, shipwrecks, forts, and lighthouses. Our exploration, though brief, will extend across the full length of the Coastal Plain, from the mouth of the Cape Fear River to Currituck Sound.

The Roanoke River and Its Namesake Refuge

If "it" in *A River Runs Through It* stands as a metaphor for the natural world, then the Roanoke River and its bottomland hardwood forests are indeed exemplars (Fig. 4-1). Some have cited this wetland ecosystem as one of the last great places in a world otherwise well laden with the footprints of human development. Even so, blemishes mar the form and function of this, the largest tract of its kind in the mid-Atlantic region. Dams alter the timing and intensity of the river's natural regime of seasonal flooding, and logging has claimed much of the cypress, leaving a preponderance of tupelo gum at many locations.

FIGURE 4-1. A National Wildlife Refuge protects tracts of bottomland forest along the Roanoke River. This type of habitat, once extensive but now limited, provides food and cover for a variety of wildlife.

The Roanoke River arises in Virginia at the eastern edge of the Appalachians and flows southeast across the Piedmont into the Coastal Plain of North Carolina, where it continues until entering Albemarle Sound. As the river approaches the northern border of North Carolina, the last of six dams impound reservoirs designed to control floods and/or to generate hydroelectric power: Kerr Lake, Lake Gaston, and Roanoke Rapids Lake. The Fall Line at Roanoke Rapids marks the beginning of the river's lower course and the development of bottomland hardwood forests. Archaeological evidence indicates that Native Americans occupied these forests 12,500 years ago and, in response to the serious spring floods, regarded the Roanoke as the "River of Death." In the mid-1600s, this region became the site of the Albemarle Settlements, the first permanent English communities in what is today North Carolina.

Below the Fall Line, rivers such as the Roanoke meander as they traverse the flat terrain of the Coastal Plain, a pattern induced by relatively small changes in slope. Meandering slows the rate of flow; hence, seasonal floods spread across the adjacent landscape, where the water-borne sediments settle out and develop rich soils—a landform geographers call a floodplain. The coarsest and heaviest sediments fall out nearest the river, where they accumulate and form natural levees. Over time, additional levees develop in response to changes in the river's hydrology, which may produce a series of ridges that parallel the channel. Conversely, floods carry the finer, lighter sediments farther inland; the materials settle at sites where the water moves with little or no force. These areas become "back swamps" that develop between the levees and the higher ground of the surrounding uplands. Within these wetlands, elevational changes of just three inches can alter the vegetation: for example, a stand of oaks may flourish on a slightly higher, drier site while a community of red maple and green ash grows nearby at a lower, wetter location. In all, the floodplain becomes a complex of forest types collectively known as bottomland hardwood forests, which often extend several miles inland on either side of the river channel.

In 1989, blocks of bottomland forest bordering the river below Roanoke Rapids were formally protected with the establishment of Roanoke River National Wildlife Refuge. Today, the refuge consists of seven units, totaling almost twenty-one thousand acres, with an overall mission to maintain and enhance the forest and its diverse fauna of fish and wildlife, which includes migratory birds (e.g., waterfowl and songbirds) and fishes (e.g., striped bass and eels). The forest itself is rich with vegetation that sorts into several communities, the largest of which consists of swamps historically dominated by tupelo gum and bald cypress, although because of logging the latter species occurs with less frequency than in the past. The swamps in turn can be subdivided based on water color: brownwater swamps, fed by alluvial waters from inland sources (e.g., the Piedmont), and blackwater swamps, whose water originates from sources on the Coastal Plain (see "Blackwater Streams: A Coastal Specialty"). In short, some are muddy, while others are not. The shrubby understory in these swamps includes Carolina water ash, sweet bay, and alders, but herbaceous ground cover is sparse.

Other communities include those adapted to the natural levees that border the river—elevated sites where sugarberry, sycamore, and green ash dominate the canopy, along with secondary species such as black walnut, sweet gum, and water hickory. Spicebush, pawpaw, and buckeye form a shrubby understory. At other sites where standing water occurs regularly but not for long periods, oaks of several species dominate the forest; these include swamp chestnut oak, laurel oak, willow oak, and a close kin of swamp red oak that botanists now identify as cherrybark oak. A large grass, giant cane, forms the understory at these (and other) sites, along with other grasses and sedges. The vegetation also features an abundance of vines, well represented by pepper vine, wild grape, Virginia creeper, poison ivy, and trumpet vine. On the highest locations—ridges where inundation occurs less often—American beech, shagbark hickory, American holly, and loblolly pine, along with shrubs such as ironwood, blueberry, and dogwood, develop what ecologists call a mesic mixed hardwood forest. (In the lexicon of ecology, "mesic" indicates a moderate amount of available moisture and thus lies between the extremes of wet and dry, respectively identified by the terms "hydric" and "xeric.")

Scattered semipermanent impoundments occur throughout the bottomland forest. These low-lying sites remain flooded with deep water long enough to prevent trees from surviving and instead develop as wetlands with herbaceous vegetation, including duck potato, picker-elweed, and smartweeds along with shrubs such as buttonbush and swamp rose.

Such diversity in the flora favors an equally rich avifauna. Indeed, more than 220 species of birds occur in the Roanoke floodplain, among them 90 breeding residents and 40 breeding neotropical migrants (i.e., birds such as warblers, vireos, and tanagers that nest in North America and overwinter in South or Middle America and the Caribbean islands). This assemblage in fact represents the greatest diversity of breeding birds known on the Coastal Plain of North Carolina. Among these, Swainson's warblers represent a species of "conservation concern" in part because of their affinity for thick understories of giant cane, but as the bottom-land hardwood forests vanished, so too have the canebrakes. Nearly all of the global population of this species—estimated at just ninety thousand birds—breed in the bottomland hardwood forests of the southeastern United States; some isolated populations breed elsewhere (e.g., moun-tain valley forests in western North Carolina). Mississippi kites, in the course of their steadily expanding range, now regularly nest in the bot-tomlands protected by the refuge. What may be the largest inland rook-ery in the state also occurs on the refuge; great blue herons, great egrets, and anhingas roost at the forty-acre site. Bottomland hardwood forests throughout much of the region likewise provided habitat for two other birds, now extinct: ivory-billed woodpeckers and Carolina parakeets (Fig. 4-2).

Whereas the Roanoke River provides American eels with a highway to aquatic habitats far inland (described in a following section), other fishes live year-round in the sluggish waters of the river's back swamps. Some regard two of these, longnose gars and bowfins, as "living fossils." Indeed, the lineage of both extends to the Jurassic Period, a time more often associated with dinosaurs than with fishes. Their appearance today differs little from the fossils of their ancestors.

Both species are rife with bones (bowfin skulls consist of twenty-eight bones organized in two layers), and both obtain oxygen in two ways: from water passing over their gills and from gulping air at the surface. The latter method clearly facilitates their ability to survive in stagnant

FIGURE 4-2. As depicted here, a male Carolina parakeet feeds on cocklebur, a favorite food of the now extinct species. Facsimile adopted from Audubon's original painting by Elizabeth D. Bolen. Photo credit: Dale Lockwood.

waters with depleted oxygen, including shallow wetlands isolated when the floodwaters recede. Both are predators well equipped with strong jaws and ample teeth; fish dominate their diets, although in some rivers bowfins prey heavily on crayfish, otherwise known as crawdads. Gars have evolved extended, narrow, and heavily toothed jaws that resemble the snouts of some crocodilians, whereas bowfins have more conventionally shaped mouths stuffed with needle-like teeth. Bowfins stealthily approach their prey by undulating their exceptionally long dorsal fins in ways that can propel the fish both backward and forward, whereas gars, also ambush predators, attack with less finesse.

River-Bottom Parrots Forever Gone

Bottomland hardwood forests have their own ghosts. Ivory-billed woodpeckers, a special focus of Alexander Wilson and probably at the northern edge of their range in the old-growth cypress swamps in southeastern North Carolina, faded away sometime in the 1940s. In contrast, the formal end for Carolina parakeets can be dated with a bit more certainty—a male named Incas died in captivity on September 21, 1918, ironically in the same cage at the Cincinnati Zoo where the last passenger pigeon, Martha, expired in 1914. Nonetheless, alleged sightings of wild parakeets occurred well into the 1930s and as late as the early 2000s for ivorybills, but these sightings lack the documentation that unequivocally proves the case (e.g., diagnostic photographs or DNA from feathers or droppings). In 1939, the American Ornithologists' Union declared that Carolina parakeets were extinct.[1]

Carolina parakeets ranged widely across the eastern United States and westward into the grasslands where suitable habitat occurred in the riparian forests along the edges of prairie streams and rivers. Their range extended as far north as New York, Wisconsin, and South Dakota, but the core of their distribution lay in the southeastern states. They were one of just two species of parrots included in the avifauna of the United States. (The range of the other, the thick-billed parrot, once included the American Southwest, but the species now occurs only in northern Mexico.) Early explorers and colonists encountered "parrokeetos" in considerable numbers along the Cape Fear River and on Roanoke Island, ample evidence of their presence in North Carolina, although, surprisingly, museums lack any specimens that were indisputably collected in the state.[2] A recent study refined the distribution of

---

1 In 2009, a news release from what seemed to be a reputable source announced the discovery of an isolated population of the vanished parakeets in Honduras and stated that details would soon appear in a leading scientific journal. The welcome news rippled through the ornithological world, but, alas, it was short-lived and no more than a sly April fool's hoax.

2 In 1963, Paul Hahn, after contacting about a thousand museums and private collectors worldwide, listed just one specimen (out of 736 skins, skeletons, and mounts of Carolina parakeets) that might have originated in North Carolina. The label attached to a specimen at the Muséum de Rouen in France states only "Carolina" and "1840s" as background data. A mounted bird at the North Carolina Museum of Natural Sciences lacks any information at all, as do a considerable number of other specimens of Carolina parakeets in the collections Hahn surveyed.

Carolina parakeets; in North Carolina, their range was limited to a narrow strip along the eastern edge of the Coastal Plain, where, as elsewhere, they were closely associated with bottomland hardwood forests, especially those with extensive swamps. In particular, groups of thirty or more parakeets roosted and nested in the cavities of mature cypress and sycamores.

Their diet consisted of a wide selection of seeds and fruits such as those of hackberry, beechnut, maple, mulberry, and weedy plants. Cocklebur seeds were a favorite, as reflected in Audubon's choice to paint several parakeets plucking apart the prickly burs on just such a plant. Carolina parakeets also foraged on crops—grains and fruit—which understandably angered farmers, who expressed their displeasure with shotguns. The gregarious nature of the parakeets heightened their vulnerability: dead and wounded birds became decoys that kept the flock circling within shooting range until most of these were also killed. Others were shot for their feathers—adornments of particular importance to milliners. Still, as tempting as it may be to blame this practice, shooting likely was not a factor contributing to the extinction of Carolina parakeets.

Cockleburs deserve further mention. The plants, including their seeds, contain toxic compounds strong enough to poison and even kill livestock, yet they posed no harm to Carolina parakeets. Still, if Mark Catesby got it right, the innards of the birds are "certain and speedy poison to cats." When Alexander Wilson tested this relationship the unsuspecting cats survived, but he added that the parakeets from which he removed the intestines had not recently eaten cockleburs. So the case for secondary poisoning remains less than "certain," and, given the extinction of the birds, cats everywhere can now breathe easier.

Ornithologists have proposed a number of causes for the extinction of Carolina parakeets. Overshooting, once generally accepted, no longer seems an adequate explanation, as it never matched the level of exploitation that spelled the end for passenger pigeons. Even more telling, Carolina parakeets winked out years after the shooting spree ended. Lumbering in bottomland forests, especially for cypress, also offers a tempting explanation, but the birds were in decline midway through the nineteenth century, well before commercial logging had removed enough trees to produce such an effect. Another idea suggests that honey bees, introduced by pioneers, preempted the nesting cavities used by the parakeets. Indeed, two other cavity nesters, purple martins and chimney swifts, also abandoned hollow trees at the same time as the parakeets started to wane, but with a difference: the swifts and martins shifted to chimneys, birdhouses, and other human structures, whereas the parakeets lacked the same adaptability. In the words of English professor and naturalist Christopher Cokinos, the parakeets "lost their homes because trees were felled [for wood] or filled [with bees]." Disease offers a plausible but unproven explanation, as parrots are unusually vulnerable to a number of pathogens, among them avian pox, avian cholera, and Newcastle disease, and barnyard poultry provides a reservoir of diseases for which native birds have little resistance. The same flocking behavior that facilitated easy shooting also would have rapidly spread a contagious disease. In all, the fatal scenario that claimed the species may never be revealed—only the feathered ghosts in dark river bottoms know the full story.

The parental care of bowfins differs significantly from that of those fishes that essentially abandon their eggs after spawning. Males build a bowl-shaped nest of aquatic plants in which one or more females will lay their eggs after a brief courtship. The males also nip off bits of plants to form a bed for the eggs inside the nest, which is often located under logs or between tree roots. After spawning, the females leave, perhaps to seek the nest of another male. On hatching, the larvae attach themselves to the nest with a sticky patch on their snouts and survive on yolk. After about nine days, and now able to swim and forage on their own, the small fish ("fry") move about in a ball-shaped school herded and guarded by the dutiful male until they reach about four inches in length—in all, a lot of parental TLC for a so-called primitive fish.

As mentioned earlier, dams miles upriver have altered environments downriver. For example, releases of impounded water in amounts that artificially alter the river's historic rate of flow over long periods of time produce severe bank erosion, known as mass wasting: large chunks of bank slump into the river along with the riparian vegetation at the site. In other words, such flows overpower the river's natural morphology. Moreover, the releases have reduced the active floodplain along the Roanoke River from its original size of 250,000 acres to about 145,000 acres, a loss of almost 60 percent. Hence, forest-dependent species must now rely on smaller patches of forest in order to sustain viable populations, which underscores the importance of the bottomland forests that remain within the refuge. Unfortunately, the abrupt transition between these patches and adjacent farmlands serve as predator lanes; as shown experimentally, the nesting success of birds breeding near these farm-forest edges diminishes when compared to the gradual transition between the two dominant types of forest vegetation (i.e., cypress-gum swamps and natural levees). Finally, the releases of water from the reservoirs do not necessarily coincide with the natural hydrological regime, an evolutionary force that has fashioned this ecosystem for millennia. Forest vegetation thus may be flooded at times when drier conditions would otherwise provide better seed germination, seedling survival, or essential aeration of root systems.

Meanwhile, logging continues in the tracts of bottomland forests held in private ownership, even on those where selective cutting years ago removed the last mature cypress. Clear-cutting operations now remove trees of virtually every species, which can be profitably processed into pellets and shipped overseas as fuel for power plants. The clock is ticking, and without a fundamental change in land-use policy, the axe and plow will relentlessly claim more acres of an irreplaceable resource. For now, Roanoke National Wildlife Refuge and about eighty thousand acres of lands held in conservation easements serve as guardians of an ecosystem in peril.

Cape Fear River and Its Dammed-Up Fish

The Cape Fear River, stretching for 202 miles, is the longest river entirely within the confines of North Carolina. Its basin, also the largest in the state, collects water from twenty-nine counties and encompasses a third of the state's population and a large share of its industrial development. Far upstream, five small populations of an endangered minnow known as the Cape Fear shiner persist in the rocky shallows of the river's headwaters. Oddly, the species features unusually long and highly convoluted intestines, which distinguish it from its nearest relatives. Dams underlie the shiners' precarious status: they interfere with the shiners' seasonal movements, but of greater importance, they flood the shallow habitat that the shiners require. Downstream, where the Cape Fear crosses the Coastal Plain, dams likewise imperil other fishes: American shad, striped bass, and two species of sturgeon (Fig. 4-3).

As the water warms each spring, American shad return to spawn in the Cape Fear River and its tributaries, as well as to the Delaware and other major river systems on the Atlantic coastline. A storied species, shad have been dubbed "the fish that fed the nation's founders," in part because legend has it that George Washington encamped at Valley Forge because of his awareness that shad would arrive in a nearby river just in time to feed his troops after a winter of short rations. True or not, water temperatures of fifty to fifty-five degrees initiate the upstream "runs" of shad returning to the same locations where they hatched years earlier. The adults rely on a keen sense of smell as their primary means of locating their natal areas, some swimming upstream a hundred miles or more to return "home." Like salmon, such a cycle identifies shad as an anadromous species, which have evolved this behavior to preclude the necessity of repeatedly searching, perhaps fruitlessly, each year for suitable spawning areas. That is, the fish return to sites with a history of successful spawning as proven by their own beginnings, as well as by those of their ancestors.

Mature females produce about 130,000 or more eggs per season; when fertilized, these hatch seven to ten days later. The fry seek sheltered areas (e.g., freshwater marshes), where they grow into fingerlings; in fall, they head downstream to the ocean, where they join others that hatched elsewhere. For the next four to five years, the young shad will mature in large schools that loop back and forth between the mid-Atlantic coast in winter and the Gulf of Maine and the Bay of Fundy in summer. As plankton feeders (they lack teeth), shad provide an important link in marine food chains for predators such as bluefish. When sexually mature, the adults return to their natal rivers, but unlike Pacific salmon, they do not

FIGURE 4-3. Anadromous fishes affected by dams on the Cape Fear River
include the endangered shortnose sturgeon, American shad, and striped bass
(*top to bottom*). Photo credits: Mary Moser, National Marine Fisheries Service
(*middle*), Sima Usvyatsov (*top*).

necessarily die after spawning. Those shad spawning in rivers north of Cape Hatteras may survive for more than one breeding season, whereas virtually all others succumb after spawning just once. Anglers prize American shad for their size (three to eight pounds), sporting qualities, and table fare, including their roe. Mysteriously, the fish eat nothing on their spawning runs, yet some will bite on lures known as darts, perhaps because of aggression rather than hunger.

In his paintings of New World fauna circa 1585, John White included a striped bass, likely a fish caught in a net set by Native people in Pamlico Sound. Seven or eight horizontal black stripes on the fish's silvery white sides and two dorsal fins, the first with nine spines, the second with just one, easily distinguish the species. Coastal anglers prize striped bass, and catches of fifteen-to-twenty-pounders prompt bragging rights. Some grow considerably larger—the current record for rod and reel weighed almost eighty-two pounds. As an anadromous species, the adults return to their natal rivers, where they spawn in the swift water at the Fall Line.[1] Up to twenty males may gather around a laying female in a splashing commotion called a "rockfight" as they fertilize her eggs. The semibuoyant eggs drift downstream for two to three days before hatching, relying on the current to keep them from settling on the bottom and becoming smothered by sediments. Initially, the fry rely on their yolk sacs for energy, but they soon feed on zooplankton. As fingerlings, they prey on small fish and later move to the ocean, where they reach breeding age in two years.

Two species of sturgeon, a group that has remained little changed since they first appeared in the fossil record more than two hundred million years ago, currently face hard times at several locations along the Atlantic coastline, including the Cape Fear River estuary. Both shortnose

---

1   In 1941, completion of a dam on the Santee River in South Carolina trapped spawning striped bass upstream. Remarkably, the isolated fish established a population that could sustain itself entirely in fresh water. The news spread, and striped bass were stocked in lakes elsewhere, including Lake Jordan in the Cape Fear River's watershed. However, successful spawning developed only in those lakes where the transplanted fish had access to fast-moving rivers, which was seldom the case. Hence, most of these populations must be continually stocked with hatchery-produced fingerlings.

and Atlantic sturgeon appear on the endangered species list. Instead of scales, five rows of boney plates (scutes) protect the bodies of these ancient fish with cartilaginous skeletons. As bottom feeders, their toothless, protrusible mouths are positioned on the underside of their heads; four sensory barbels under their snouts detect mollusks, crustaceans, and other benthic foods. Sturgeon also locate their prey with sensory organs known as ampullae of Lorenzini—small pits on their heads in which nerve endings can pick up the electrical currents emitted when animals move their muscles, an adaptation they share with sharks. In general, they feed indiscriminately and simply vacuum up whatever foods they encounter. Curiously, some species leap entirely out of water, a performance typically occurring at dawn or dusk and only in fresh water. Like the breaching of humpback whales, the feat lacks a satisfactory explanation, although some evidence suggests it may be a means of communication associated with group cohesion. Sturgeons have long been heavily exploited for meat and, particularly, for their roe—the delicacy known as caviar.

Shortnose sturgeon are particularly rare in the Cape Fear and other rivers in North Carolina. They spawn in fresh water, where the fry remain for up to five years before moving downstream to the tidal zone, where fresh and salt water meet. When mature, the adults reach lengths of three to four feet and weigh up to fifty pounds. Shortnose sturgeon often ingest considerable amounts of mud along with their food, and they may occasionally feed on snails attached to aquatic vegetation. Spawning runs in North Carolina begin in early spring to sites with hard bottoms; the eggs, once fertilized, adhere to the rocks or other submerged objects. After spawning, the adults return downstream and spend the remainder of the year in the estuary or in the nearby ocean. Shortnose sturgeon prefer water of less than 3 percent salinity, and some survive entirely in freshwater environments. They seldom venture far from their natal rivers, which precludes contact between populations in adjacent estuaries.

As adults, Atlantic sturgeon weigh in at five hundred to eight hundred pounds and reach fifteen feet in length. Like shortnose sturgeon, they spawn in fresh water but otherwise spend their lives in salt water, including the ocean, where they may travel along the coastline for several

FIGURE 4-4. Terraces of rocks known as a fishway constructed on the downriver side of Lock and Dam 1 on the Cape Fear River help anadromous fishes reach their spawning grounds farther upstream.

hundred miles before returning to their natal rivers. Their spawning runs occur several weeks after those of shortnose sturgeon, which likely reduces competition between the two species for spawning habitat. Atlantic sturgeon prefer to spawn in deep pools with a strong current, often at the base of rapids, but such habitat is not a natural feature of the Cape Fear River below the Fall Line. Some females may spawn every year, but others spawn at intervals of three or more years. Although adults of both species share a similar diet, competition for food diminishes because the fish occupy different habitats for most of the year (i.e., sites partitioned by salinity).

As mentioned earlier, the Cape Fear River differs from other streams in the state—three navigational lock and dam units, each about thirteen feet tall, span the river on the Coastal Plain. In contrast, dams on rivers such as the Roanoke occur in the Piedmont beyond the Fall Line. The first lock and dam on the Cape Fear was built in 1915 about sixty channel miles from the river's mouth, followed by similar structures farther upstream in 1917 and 1934; none was constructed with fish ladders, and the migrant fish populations accordingly diminished. Beginning in 1962, the locks were managed to allow the fish to enter and move upstream or downstream in the same manner as the barges, but the effort provided limited results. In 2012, a fishway—a ramp of rocks in a series of step-like terraces—was constructed along the downriver face of Lock and Dam 1 so that migrating fish might ascend from the river, swim over the dam, and continue upstream (Fig. 4-4). The fishway worked well for shad but much less so for striped bass, although a proposal to widen the gaps between the larger rocks so that schools of striped bass can move together up the fishway may provide a remedy. If the project proves successful, it will serve as a model for similar modifications at the other two dams. However, whether these modifications will help sturgeon again reach their historic spawning areas remains unclear.

Blackwater Streams: A Coastal Specialty

Oceangoing ships of days long gone often carried barrels of tea-colored water obtained from watersheds confined to the Coastal Plain. While seemingly unsavory, the dark water is also acidic, which checked the proliferation of microbes, so the water remained "sweet" or potable even after weeks at sea. Crews and passengers alike safely drank the stained water, perhaps improved at times by a wee dram of rum.

These same properties—staining and acidity—shape the aquatic communities in what are aptly known as blackwater streams. The staining originates from tannins and dissolved organic matter leached from terrestrial vegetation and thereafter washes into the streams, a happenstance accentuated on sandy soils. More to the point, the stain limits the penetration of sunlight in the water column, which diminishes the diversity of aquatic vegetation in blackwater streams. Nonetheless, the flora may include beds of species such as bur reed, wild celery, and ribbonleaf pondweed, but even these generally occupy relatively small areas of the streams. Special mention goes to Cape Fear spatterdock, which is closely associated with blackwater environments in the Coastal Plain (Fig. 4-5). Instead of oval or disc-shaped leaves like those of its relatives, this species bears narrow floating leaves with wrinkly edges; its thin

FIGURE 4-5. Cape Fear spatterdock, primarily associated with
blackwater streams, features elongated leaves with wrinkled margins
quite unlike the rounded leaves of its relatives.

and lettuce-like submersed leaves are likewise distinctive. In season, the
plants produce green or yellow cup-like flowers atop stalks that emerge
about ten inches above the water. The dark waters also lessen the density
of algal populations, which, in turn, diminishes the density of copepods
and other zooplankton. In all, only a limited amount of primary production—the photosynthetic creation of organic matter—arises *within*
blackwater streams. Instead, much of the energy that fuels life in blackwater streams originates from the decomposition of leaves and woody
debris imported from the surrounding floodplain.

Contrary to popular opinion, blackwater streams support a rich diversity of invertebrates, although not all groups are equally represented in
comparison with other types of streams. For example, dragonflies and
damselflies of several taxa in blackwater streams often outnumber those
in alluvial streams, as do dipterans; of the latter, blackflies and midges

occur in abundance. On the other hand, blackwater streams support only a few species of freshwater mollusks—largely snails and mussels—because of the limited supply of calcium carbonate in the acidic water. Fingernail clams, however, have a greater than average tolerance to acidity and may be abundant in some blackwater streams.

Fishes closely associated with blackwater streams include the blackbanded sunfish. Well named, these small fish, which reach the size of a silver dollar, have six vertical black lines on each side and a black spot on their gill covers. In keeping with their distinctive markings, blackbanded sunfish occur in dense beds of aquatic vegetation, including filamentous algae. They cope with acidities as low as pH 3.7 by a means as yet undetermined but likely one that involves the physiology of their gill tissues. Conservation agencies in North Carolina consider the blackbanded sunfish as a vulnerable species because of its narrow niche and patchy distribution and threats from habitat degradation. Redfin pickerel also occur widely in blackwater streams, as do a number of other popular game species (e.g., bluegills and largemouth bass). Numerous species of forage fishes—dace, darters, and shiners—provide a diverse food web. The diets of many fishes in blackwater streams also include terrestrial organisms originating in the adjacent floodplains, and indeed many fishes forage in these areas whenever they flood. Snags and other woody debris in the channels provide important habitat for aquatic insects, among them mayflies, that further enrich the prey base for fishes in blackwater streams.

Because of their low-gradient profiles as they cross the Coastal Plain, blackwater streams have escaped hydroelectric development, and most have maintained their original character. Indeed, the ecological and aesthetic attractiveness of blackwater rivers has led to their recognition and protection as nature preserves, outstanding waters, and scenic rivers. As an example of the latter, the Lumber River, considered one of North Carolina's Ten Natural Wonders, was designated as a National Canoe Trail in 1981 and as a National Wild and Scenic River in 1998. More than just appealing, the Lumber River also provides habitat for an exceptional biological community, including rare species such as sarvis holly and the sandhills chub. Some 11,200 acres of this outdoor treasure lie within the bounds of Lumbee River State Park.

Ancient Trees on the Black River

The course of the Black River, so named for its tannin-stained waters, travels entirely within the Coastal Plain, where it flows for sixty miles before merging with the Cape Fear River. Although dark, the waters nonetheless are of high quality and warrant the state's highest ranking as an Outstanding Resource Water. The presence of several species of freshwater mussels, some rare, further reflects the purity of the Black River; these include the Atlantic pigtoe, pod lance, and Cape Fear spike, all intolerant of water of diminished quality.

But the Black River offers more than clean water and rare mussels. Indeed, a river-edge site known as Three Sisters Swamp harbors the oldest trees east of the Rocky Mountains, one of which, aptly named Methuselah, started from a seed germinated in the fourth century—and some may be older yet! Some of these trees, all bald cypress, exceed eight feet in diameter (Fig. 4-6). Unfortunately, rot has hollowed many of their trunks, thereby precluding a full accounting of their true age (i.e., annual growth rings no longer remain deep inside the trunk). Still, one revealed 2,624 rings, by far the oldest so far sampled at the site and the oldest living tree in eastern North America. Indeed, these ancient trees prompt their portrayal as the "redwoods of the East." Centuries of storms have torn away their upper branches, leaving these old-timers with flattened crowns. These trees escaped the logger's axe because of the difficulty of removing logs from the swampy terrain. They stand today festooned with Spanish moss, which is not a moss; as unlikely as it may seem, the flowering parts of the species show its association with the pineapple family! Note that, unlike parasitic mistletoe, Spanish moss is an epiphyte that acquires its nutrients from airborne moisture and relies on trees solely for support. Some bats commonly roost in the plants' tangled beards. The distribution of one such species, the Seminole bat, in fact closely coincides with the range of Spanish moss. Where available, Spanish moss likewise conceals the nests of northern parulas—small migratory warblers—although the nest itself may consist of other materials. A harmless species of jumping spider also resides, perhaps exclusively, in the plants, but contrary to popular belief, mite larvae known as chiggers

FIGURE 4-6.
Ancient stands
of bald cypress,
including some
more than two
thousand years
old, survive along
the Black River,
aptly named for
its tannin-stained
water.

or red bugs do not. Beware, however, as these skin-piercing and highly irritating larvae indeed occur in clumps of Spanish moss that happen to lie on the ground.

The ancient trees offer prime roosting habitat for Rafinesque's big-eared bat, which is listed as a species of concern throughout its range (Fig. 4-7).[2] Apparently foraging almost exclusively on moths just a few feet above the ground, these agile fliers can hover as they search for hapless victims—a behavior unlike that of many other bats that catch their prey in full flight well above the ground. Rafinesque's big-eared bats occur in scattered populations in the southeastern United States as far west as eastern Texas. Within this range, they use caves but favor mature stands of bottomland hardwood forest near permanent water—a habitat type that itself is no longer commonplace in the Coastal Plain of the Carolinas or anywhere else in the southeastern states.

Rafinesque's big-eared bats do not migrate and instead remain in the same vicinity year-round. On cool days, they enter torpor for a few hours to save energy but will hibernate for weeks or even months during prolonged periods of cold weather. The species remains faithful to roosting sites in a relatively small area of the forest and cling to both vertical and ceiling surfaces within the cavities. Although unproven, it seems likely that the loss of roosting sites may lie at the core of the species' rare status. Moreover, disturbances may cause bats to abandon regularly used roosts, and given the scarcity of large trees with suitable cavities, this factor gains additional significance when it concerns colonies of females raising their pups.

Rafinesque's big-eared bats may possibly be at risk from a fungal disease known as white-nose syndrome (WNS), which has already claimed

2   Naturalists of all stripes owe it to themselves to become acquainted with Constantine S. Rafinesque (1783–1840), a Turk by birth and an eccentric genius by nature who spent much of his life describing—by the hundreds!—plants and animals in North America. While visiting John James Audubon (1785–1851), he wielded the painter's violin to swat at bats that he thought might represent undescribed species but, alas, were not. In turn, Audubon drew several fanciful animals that he represented to Rafinesque as new species. As hoped, his eager guest took the bait and unwittingly described a few of these, including a fish with "bullet-proof scales," in a ruse that remained undiscovered for several years. See Warren (2004).

FIGURE 4-7. Rafinesque's big-eared bats often roost in hollow trees near water. Agile fliers, these bats also can hover when feeding on moths and other insects.

millions of bats of at least seven species in North America. True to its name, the fuzzy white fungus develops on the nose of infected bats to the extent of rousing them from hibernation. The hungry bats then go in search of food, but given the absence of insects in winter, they soon die from starvation. WNS is highly contagious and quickly spreads to new areas and to other species that share the same winter roosts in caves. However, WNS so far has not affected tree-roosting populations of Rafinesque's big-eared bats. Note that insectivorous bats of all species collectively provide an important ecological service—when active, they literally consume countless tons of insects every night. Accordingly, economic research suggests that their impact on harmful insects yields a value to the agricultural industry of about $22.9 *billion* per year, which includes not only increased crop production but also savings from the reduced need for pesticide applications.

Ancient stands of bald cypress may reveal crucial events related to the conservation of any ivory-billed woodpeckers that might miraculously still survive in North America, as was reported in 2005. When lengthy droughts occur, as shown by climatic reconstructions based on tree-ring data, they create ideal conditions for the regeneration of bald cypress. The trees produce viable seeds throughout their long lives, but the seeds will not germinate in water, nor will the seedlings survive prolonged inundation. Hence, droughts become an event necessary to establish each new generation of bald cypress. At the same time, trees dying from these extensive droughts become habitat for the larvae of wood-boring insects—the predominant food for ivory-billed (and other) woodpeckers. Droughts thus produce new stands of bald cypress while concurrently increasing the food resources required by the rare birds. In short, preservation of these ancient trees and the history of drought they reveal may be of crucial importance to restoring ivory-billed woodpeckers—that is, if any can be found as stock for reintroduction.

Drought may also have caused the downfall of the Lost Colony on Roanoke Island. Tree-ring chronologies from ancient bald cypress growing on the Coastal Plain indicate that an extended dry period occurred at the very time the settlement failed. In fact, a three-year drought between 1587 and 1589 was the most extreme event of its kind during the eight-hundred-year period covered by the chronologies. Similarly, a long drought coincided with the increased mortality at the colony established at Jamestown, Virginia. Unlike Roanoke, Jamestown did not fail, but some forty-eight hundred of the six thousand settlers arriving there between 1609 and 1625 perished in the midst of a lengthy drought. These hard times not only affected crops and livestock but also likely fouled water quality (i.e., increased salinity in wells and nearby estuaries). Had it not been for droughts, life in these English settlements might have been far less severe—a story now told by the growth rings of ancient trees.

Fortunately, The Nature Conservancy now protects thirteen miles of critical habitat along the course of the Black River. The protected area includes more than three thousand acres of old-growth cypress designated as the Black River Preserve, which, not incidentally, also provides a refuge for Rafinesque's big-eared bat along with a host of other wildlife. Moreover, another layer of protection may be forthcoming. In 2017, the governor of North Carolina signed a bill authorizing a feasibility study as a first step in designating the site as a new state park.

Another Lost Colony?

History books relate the saga of Virginia Dare and the "Lost Colony" of Roanoke Island, which vanished into the mists of history sometime after 1587 (see "Ancient Trees on the Black River"). But just as mysterious, at least for those interested in birds, is the irregular disappearance of a large nesting colony of white ibis from an island in the mouth of the Cape Fear River, an event that occurred most recently in 2018. Normally, five thousand or more pairs of white ibis, which represents about 10 percent of the North American population, nest at Battery Island each year (Fig. 4-8). Such numbers signify a colony of global significance and thus receive rigorous protection from wardens working for North Carolina Audubon, backed in full by legal safeguards from state and federal laws. Large ibis colonies also exist in the Everglades and elsewhere, and they too at times suddenly and inexplicably disappear.

Battery Island covers about one hundred acres, and except for the addition of some dredged sand years ago along its southern edge, the island formed naturally from deposits of marine sediments. Coastal red cedar and yaupon characterize the woody vegetation that white ibis and several other species of wading birds (e.g., herons) select as nesting sites. Each of these species respectively builds its disheveled-looking stick nests at different heights above ground—a partitioning that helps reduce interspecific competition. White ibis typically lay clutches of two to four eggs, which hatch after twenty-two days of incubation; the parents then provide their nestlings with forty to fifty more days of parental care, which includes securing food that they digest and regurgitate to nourish their broods. Oddly, the young birds do not develop functional salt glands for five or six weeks after hatching, which means they cannot survive on fiddler crabs and other marine prey gathered by their parents from the saline waters that surround Battery Island. Instead, the adults make daily

FIGURE 4-8. A large colony of white ibis usually nests on Battery Island in the Cape Fear River estuary. However, at irregular intervals and for unknown reasons, the birds abandon the site and temporarily relocate elsewhere.

flights to freshwater wetlands in search of suitable foods. These locations may be far removed from the colony, including Lake Waccamaw, some forty miles distant. Others forage for insects and other invertebrates in cultivated fields well inland, and they do so without harming the crops. Some tamely seek food on lawns and roadsides.

Just what triggers the birds to abandon Battery Island remains a mystery, but whatever it is seems to be in force for just one year at a time. Based on censuses that date to 1975, the colony has disappeared at times in the past but reestablished itself on the island the following year. The appearance of great horned owls has been suggested as a possible cause, but the threat posed by these or other predators was not enough to scare off the other species of wading birds nesting on the island. Additionally, when the ibis colony moves to a new site elsewhere, it sometimes may be difficult to locate—just like the one so long ago at Roanoke Island.

Eels: A Slippery Travel Story

American eels begin and end life in the Sargasso Sea but may spend a decade or longer maturing in freshwater environments far inland. This form of migration, known as catadromy, differs from the better-known anadromy of species such as salmon and striped bass that swim up freshwater rivers to spawn after maturing in the ocean. In fact, eels stand alone as the only catadromous fish in North America. In North Carolina, eels migrate in all of the major rivers and their many tributaries, as well as in many smaller streams and creeks that flow directly into sounds or the ocean. Because they can breathe through their skin, eels also slither snake-like overland from streams to reach isolated lakes and ponds and even swamps. Eels migrating up rivers in South Carolina once reached the Catawba River and other streams in western North Carolina, although dams and other obstacles have largely eliminated these populations. In comparison, American eels remain well established in the lower watersheds of rivers in the Coastal Plain.

The Sargasso Sea itself is unique—its waters touch no coastlines. Instead, currents such as the Gulf Stream define its boundaries. Because of its exceptionally calm waters, the Sargasso Sea was once the bane of wind-powered ships and is still avoided by those fearful of the mythical "Bermuda Triangle." Mats of brown algae—Sargasso seaweed—float on the surface, held up by bulbous bladders that to Portuguese sailors resembled grapes and subsequently identified this region of the Atlantic Ocean. European eels also spawn in the Sargasso Sea, yet both species maintain their own genetic identities.

After hatching, eels develop through several stages, the first of which is a larval form known as leptocephali. These larvae resemble flattened, transparent willow leaves with small heads and large teeth. They drift west and north for the next seven to twelve months, guided by the surface currents of the Gulf Stream. The larvae gradually move closer to shore and metamorphose into the second stage of development: glass eels, still transparent but now two to three inches long and shaped like adults. At this point glass eels move into estuaries along the Atlantic coast—Pamlico and Albemarle Sounds, for example—and begin losing their transparency; now called elvers, many start moving upstream, although some will stay and mature in estuaries. As they gain additional pigmentation elvers become known as yellow eels and acquire their sexual identity. Whereas it still remains something of a biological mystery, sexual differentiation in eels seems to be influenced by the density of their local populations: greater density results in more males, which often is the case in estuaries, as opposed to the development of more females in freshwater habitats well inland.

The final stage occurs when yellow eels metamorphose into silver eels, although their sexual organs will not completely mature until years later when they return to spawn in the Sargasso Sea. At full maturity, eels are at least ten years old, although some may be two or even three times older. They do eat on the return trip, and, in fact, their digestive tracts gradually degenerate, heralding their approaching death. Atlantic eels and Pacific salmon thus share a common fate: they complete just one migratory cycle, which ends with spawning and death. However, no one has yet witnessed American eels in the process of courtship or spawning in the wild, likely because both events occur in deep water (whereas, as commonly seen in nature programs, salmon spawn in water barely deep enough for swimming).

American eels face several threats. Dams inhibit or completely block their upriver migration, and those moving downstream to spawn often die in the whirling blades of turbines at hydroelectric plants. Eelways erected at some dams, among them one on the Roanoke River, now lessen these impairments, and similar projects are scheduled for others, but dams elsewhere still present serious physical obstacles for migrating eels. Glass eels are protected in many states, including North Carolina, whereas yellow and silver eels may be legally harvested (e.g., yellow eels make excellent bait for striped bass). However, because glass eels command huge prices as breeding stock for aquacultural ventures in Asia, they remain potential targets for poachers. A parasitic worm that infects the swim bladders of eels also places American eels at risk. The worms, native to Japan, spread to Europe, where they infected European eels and later reached North America—an intercontinental journey almost certainly enabled by the commercial trade of eels rather than by a natural means of dispersal. First discovered in South Carolina in 1995, the parasitic worms now occur widely in North Carolina's rivers and sounds, as well as in other locations along the Atlantic coast. The infections, although not necessarily fatal, likely impair the buoyancy and swimming performance of adult eels, as shown experimentally in European eels infected with the same parasite. If so, infected eels may not survive the long spawning run back to the Sargasso Sea. Because of these and perhaps other factors, American eel populations have declined at alarming rates in recent decades, and two attempts to place them under the protection of the Endangered Species Act have failed.

A Snail's Slide toward Extinction?

In 1903, naturalist Henry Pilsbry described a new species of snail with a large and distinctively shaped shell; he identified the collection site only as "lower Cape Fear River region." With this brief introduction, the magnificent ramshorn snail was formally added to the aquatic fauna of North Carolina's Coastal Plain (Fig. 4-9). The species turned up again five years later in Greenfield Lake, now within the city limits of Wilmington, but thereafter remained obscure until 1988, when it was discovered some

FIGURE 4-9. The endemic and extremely rare magnificent ramshorn currently exists only in captivity; with successful breeding, wild populations may be reestablished in southeastern North Carolina. *Inset*: fresh egg mass on the leaf bottom of a yellow water lily, a plant important to the snail's life cycle. Photo credit: Andy Wood, Coastal Plain Conservation Group.

fifteen miles away in a privately owned lake in Brunswick County. In 1992, the species was discovered in a Brunswick County millpond that connects with the lower Cape Fear River by way of a tributary stream. The most recent and last discovery of the species in the wild occurred in 2004 in McKenzie Creek, a blackwater stream in the lower Cape Fear River basin. All of these sites lie within a forty-square-mile area, which strongly suggests that magnificent ramshorn snails, while perhaps more widely distributed in the past, today persist in the wild—if at all—in a range far more restricted than the well-known Venus flytrap. None has been seen in the wild since 2004, despite repeated searches within the area, including those sites where they previously occurred. Magnificent ramshorn snails, notwithstanding their pretentious name, lack the appeal of those "charismatic species" (e.g., giant pandas and whooping cranes) that capture the public's compassion or the attention of conservation agencies.

Magnificent ramshorn snails at maturity reach the size of a quarter; they are the largest species in the family Planorbidae, a group that breathes with lungs, not gills. Their shells coil to the right—yes, snails separate into species that turn in one direction only, right or left—and lie flat like a coil of rope. They lack an operculum, the scale-like structure that plugs the opening after they withdraw into their shells. The opening itself slants downward to the right and features a flared outer lip. Interestingly, planorbid snails transport oxygen in iron-based hemoglobin (red blood), whereas all other freshwater snails have copper-based hemocyanin (blue blood). Just why this difference emerged remains uncertain, especially since early in their evolutionary history planorbids were also blue-blooded. Early reports claim that the snails lose their eyes as they mature, but, if true, that anomaly also lacks an explanation.

Like many other snails, magnificent ramshorns graze on the fuzzy film of algae growing on aquatic vegetation, typically on the underside of waterlilies and other plants with floating leaves. They likely feed on detritus as well, but their disappearance has precluded field studies of their life history. Other than their occurrence in still or slow-moving waters where acidity is not prohibitive, little else can be said at present about their habitat requirements. Other planorbids serve as intermediate hosts in the life cycle of some parasitic worms, but this too is among the many unknowns about magnificent ramshorns.

Fortunately, the species survives in captivity. Naturalist Andy Wood maintains what may be the only remaining population in his aptly named Planorbella Conservatory, a battery of three-hundred-gallon plant-filled aquaculture tanks carefully managed for temperature and water quality. The immediate goal, of course, is to prevent the extinction of the species but ultimately to produce enough snails for release into the wild. Like many other snails, magnificent ramshorns are hermaphrodites; hence, the possibility of an imbalanced sex ratio in a small breeding population is of no concern—any two individuals, if willing and able, can mate. Even with adequate stock, however, suitable habitat for releases may be hard to find. Some sites have been choked by invasive noxious vegetation, others have been treated with herbicides, and still others have become too saline for the salt-intolerant snails. In the latter case, extensive dredging of the ship channel in the lower Cape Fear River has enabled salt water to encroach miles upstream from the river's mouth. As a result, salt water also penetrates the smaller creeks in the drainage, which is exactly the region cited in Pilsbry's original paper.

Saving any species from extinction, no matter how insignificant it may seem, remains a priority for conservationists everywhere, and the magnificent ramshorn snail is no exception. It is a natural component of the state's fauna, a part of an aquatic ecosystem, and a treasure of nature. The species hangs on, but only barely, and although difficult, its recovery deserves no less than our full attention.

The Coastal Plain's Own Salamander

Only a mother salamander—or a dedicated herpetologist—could love a waterdog. Still, the Neuse River waterdog, albeit "unloved" by many, has the virtue of representing another unique aspect of North Carolina's Coastal Plain (Fig. 4-10). In 1924, C. S. Brimley described what he then thought was a subspecies of the common mudpuppy, but comparative studies decades later of skin patterns of juveniles and tissue proteins recognized the taxon as a full species. Its distribution extends into the Piedmont on the upper reaches of the Tar and Neuse Rivers, but fully two-thirds of the remainder lies entirely within the confines of the Coastal Plain; the species occurs nowhere else. The animals' range extends downstream to the point where salt water dominates the two rivers.

Interestingly, the irregular distribution of the species in the Trent River, which originates in the Coastal Plain and empties into the Neuse River, seems associated with outcrops of limestone, whereas elsewhere in the Trent River system they are replaced by dwarf waterdogs. Just how the limestone favors Neuse River waterdogs remains unclear, but it may be linked to the higher quality of the groundwater flowing into the riverbed where the outcrops occur (e.g., greater oxygen content and less acidic).

Neuse River waterdogs, which reach lengths of eleven inches, complete their entire life cycle in water; accordingly, the adults retain larval features such as gills, a tail fin, and no eyelids—a developmental pattern zoologists call neoteny. They eat just about anything they can catch; hence, their diet includes insects, snails, leeches, small crustaceans, and, occasionally, small fish, as well as terrestrial species such as earthworms, spiders, and centipedes, among others. Anglers sometimes catch waterdogs, which reflects their attraction to worms. Although they

FIGURE 4-10.
Neuse River
waterdogs occur
only in the main
channels and
larger tributaries
of the Neuse and
Tar Rivers. Photo
credit: Jeff Beane,
North Carolina
Museum of Natural
Sciences.

are opportunistic feeders, their diet does not often include prey that, although relatively abundant, may be too elusive to capture (e.g., crayfish).

In April and May, females lay clutches of nineteen to thirty-six eggs—fewer than a dozen nests have been studied—on the underside of flat rocks in water about a foot deep; females attend the nest until the larvae hatch in July. They remain active year-round within a temperature range of thirty-two to sixty-four degrees. Unlike some other amphibians, there is no clear evidence that the skin of Neuse River waterdogs produces toxic secretions that repel predators. Unfortunately, most of the field studies of the species have been conducted above the Fall Line where rocks on the stream bottom provide cover for nests and shelter, whereas other structures—bottom debris, roots, and bank features—may fulfill these functions on the Coastal Plain.

The limited distribution and uncertain status of the population warrant designation of the Neuse River waterdog as species of special concern in North Carolina—a precursor to the more serious designations of threatened and endangered (see the Afterword). Additionally, as an aquatic organism, the species faces the potential threat of water pollution, especially in watersheds subject to growing commercial and residential development.

Sounds: North Carolina's Coastal Seas

The seaward border of the Coastal Plain features a series of coastal lagoons, more familiarly known as sounds, which vary in size from the vastness of Pamlico Sound to the confines of Bogue Sound. Regardless of size, however, they all share two characteristics: shallow depths and the protection of an elongated barrier beach, whether island or peninsula. A major system consisting of Pamlico, Albemarle, and Currituck Sounds lies east of the Suffolk Scarp, a remnant of an ancient shoreline that formed during the warm Sangamon interglacial period some 125,000 to 75,000 years ago. The Talbot Terrace rises to the west of the scarp, whereas the low-lying, poorly drained Pamlico Terrace extends eastward to the coast. Although about seventy miles inland from Oregon Inlet, the scarp rises just twenty-five feet or so above the current sea level. In places, it serves as a firm foundation for state and county roads.

Pamlico Sound, the largest of an interconnected system that includes Albemarle and Currituck Sounds to the north and Core Sound to the south, is also the largest of its kind on the eastern coast of the United States (in all, eight sounds occur along the North Carolina coast). Indeed, when explorer Giovanni da Verrazzano (1485–1528) sailed into Pamlico Sound in 1524 he believed it was an entrance to the Pacific Ocean. His error becomes forgivable, given that the sound covers an area eighty miles in length and as much as twenty miles in width—far enough in either direction to extend beyond the sight of land.

Seawater enters Pamlico Sound through three inlets, Oregon, Hatteras, and Ocracoke, whereas two major rivers, the Neuse and the Pamlico-Tar, and a number of lesser streams feed fresh water into the system. Brackish water flowing from the two northern sounds also enters Pamlico Sound, and the resulting mix dilutes its salinity to about 2 percent, as opposed to 3.5 percent for full-strength seawater. Because extensive amounts of fresh and salt water meet in these sounds, they collectively represent an estuarine system second in size only to Chesapeake Bay.

The shallow water in Pamlico Sound, just five to six feet on average, supports both game and commercial fisheries, including prized species such as red drum, speckled trout, flounders, bluefish, and striped bass,

Red Tide Pays a Visit

In the winter of 1987–88, offshore currents developed conditions for a "perfect storm" along much of the North Carolina coast—not a hurricane or nor'easter but an immense algal bloom known familiarly as a red tide. Marine scientists prefer to identify these events as harmful algal blooms (HAB), largely because they are not always red, nor are they related to tidal movements. Nonetheless, the designation red tide remains well established in the public vocabulary and thus fulfills our needs.

Typically, these irruptions occur in the Gulf of Mexico, often along the west coast of Florida, where the dinoflagellate *Karenia brevis* releases neurotoxins that cause fish kills and, in some cases, the deaths of birds, sea turtles, and marine mammals.[1] (Alas, scientists do not bless single-celled algae with common names.) The toxins can also accumulate in the tissues of living fish and seagrasses, which, as food for dolphins and manatees, can prove fatal long after the bloom has dissipated (i.e., a delayed response, including mortality occurring in remote locations). Humans, too, may suffer from red tides when waves, wind, and motor boats produce toxin-laced aerosols, resulting in eye, nose, and throat irritations and difficulty breathing. One of the most common effects is the closure of waters to shellfishing in order to prevent humans from consuming contaminated clams and oysters, which can cause illness or even death. Indeed, about 571 square miles of North Carolina waters were

---

1   As opposed to "eruptions," biologists identify large and sudden increases in plant and animal populations as "irruptions," a term now widely recognized in dictionaries.

---

as well as crabs, clams, and oysters. Historically, the sounds have produced about 90 percent of North Carolina's commercial seafood catch.

In 2015, "Katherine," a fourteen-foot great white shark, briefly visited Pamlico Sound, an event heralded in newspaper headlines. Because the twenty-three-hundred-pound shark had been tagged with a transmitter at Cape Cod in 2013, biologists associated with the OCEARCH Shark Tracking Program could trace her inshore movements and, to the relief of many, document her departure eighteen hours later. Katherine, closed as a result of the 1987 bloom. Even so, the bloom caused at least forty-eight cases of neurotoxic shellfish poisoning. All told, the event produced an economic loss of more than $24 million.

Unusual oceanographic circumstances triggered the expatriate bloom. These began when an offshoot of the Gulf Stream known as the Loop Current reached the western coast of Florida coincident with a bloom in the same location. There the Loop Current picked up and transported water laden with *K. brevis* southward, where it moved around the tip of Florida—the flow at this point is known as the Florida Current—and rejoined the Gulf Stream.

The Gulf Stream continued its northward flow until mid-October, when it inexplicably meandered westward over the continental shelf at Raleigh Bay (i.e., between Cape Hatteras and Cape Lookout). The meander lasted until the end of December, and with it, *K. brevis* came inshore, where surface winds carried the bloom inside the barrier islands north of Cape Lookout. In all, the bloom persisted for four to five months (to late February 1988) and, driven by northerly winds, extended southward to Cape Fear and on to Little River Inlet in South Carolina.

The 1987–88 event represented the first record of *K. brevis* north of Jacksonville, Florida. A year later, just after another bloom occurred on the southwestern coast of Florida, a brief meander in the Gulf Stream again carried *K. brevis* to North Carolina. This time, however, the event was short-lived and gone within a week. Thus, it seems clear that the Gulf Stream and its unpredictable meanderings can at times transport red tides from the west coast of Florida to North Carolina and may do so more often than once believed.

presumably in search of food, apparently entered the sound through Oregon Inlet and returned to the ocean through Hatteras Inlet. Another great white shark, "Cabot," close to ten feet in length and likewise tagged by Ocearch, entered Pamlico Sound in November 2019. Bull sharks also gained attention by adding Pamlico Sound to the list of sites along the eastern coastline that serve as nursery areas for their young. Rising water temperatures in the sound seems to be the primary driver in this new relationship.

Albemarle Sound lies on an east–west axis, extending fifty-six miles from the inner shoreline of the Outer Banks to the current mouth of the Roanoke River; it varies from three to twelve miles in width. The linear form of Albemarle Sound reflects its origin as a drowned section of the river's lower valley; this occurred when the last ice age ended and the glacial meltwater slowly raised sea levels along coastlines worldwide. Lunar tides exert little effect in this and the adjacent sounds; hence, the wetlands bordering these systems develop as irregularly flooded "high marsh" dominated by black needlerush; swamp forests of bald cypress and tupelo gum also rim the mainland edges of these sounds. Instead of regular tides, wind acts as the primary water-moving force, particularly in Albemarle Sound because of its long fetch (i.e., unfettered wave movement, in this case for nearly sixty miles).

Fresh water flowing from the Chowan and Roanoke Rivers, supplemented by inflow from Currituck Sound (see below), lessens the salinity in Albemarle Sound. Moreover, the sound lacks direct connections with the ocean (inlets), which obviously precludes the influx of seawater along with other sorts of marine influences, likely including some hefty sharks. Together, these factors drop the salinity in Albemarle Sound to about 0.5 percent, or about 14 percent the strength of seawater. The low salinity also keeps out oysters, except during periods when prolonged droughts reduce the inflow of fresh water. The change from high to relatively low salinity at Manteo, a narrow gap separating Pamlico from Albemarle Sound, creates conditions favorable for the growth of widgeongrass north of this point.

Currituck Sound parallels the north–south orientation of the Outer Banks. Much smaller than either Pamlico or Albemarle Sound, it extends some thirty-eight miles in length and just three to eight miles in width. The sound lacks inlets, the last of which closed in the 1700s, so its waters today range from slightly brackish to entirely fresh. Oregon Inlet, some twenty-five miles to the south, remains the nearest source of seawater. Unlike its neighbors, Currituck Sound encloses a few small wooded islands. One of these, Monkey Island—the name stems from a corruption of Pamunkey, a local Indian tribe—houses a large rookery of herons and other wading birds. Shell middens (see "Oysters: More Than

Just Good Eating" below) serve as reminders of bygone days when the Indians used the island as a hunting camp. Centuries later, wealthy duck hunters also visited the island, likewise leaving behind some telltale evidence: the ruins of their hunting lodge, one of many once active on the shores of Currituck Island. Monkey Island also marks the northern limit in the distribution of dwarf palmettos. Also known as dwarf palms (they seldom exceed three feet in height), this species is one of the most cold tolerant of their kind. Monkey Island is now part of Currituck Island National Wildlife Refuge.

Submersed aquatic vegetation flourishes in these sounds. Two seagrasses, eelgrass and shoalgrass, dominated the saline areas, whereas wild celery and bushy pondweed occupied the freshwater habitat at the northern end of Currituck Sound (see below).[3] A felt-like epiphytic community coats the healthy leaves of eelgrass and other seagrasses. This community includes algae and bacteria, as well as a menagerie of minute animals that form their own food chains of predators and prey. Worldwide, hundreds of species have been cataloged on the leaves of eelgrass, and these in turn provide food for larger animals. Snails, for example, graze on the coating of epiphytes with a rasping tongue-like structure known as a radula, and small fishes of several kinds pick at the fuzz-covered leaf surfaces in search of food. The epiphytic community, if unusually heavy, may inhibit the photosynthetic activities of its host, but as the older leaves drop off—a normal event accompanied by the appearance of new foliage—the situation becomes largely self-correcting.

Eelgrass nearly vanished from the North American (and other) coasts in the 1930s when an infectious, host-specific slime mold known only by its Latin name, *Labyrinthula zosterae*, attacked the plants. The pathogen

3   Seagrasses should not be confused with the true grasses in the family Poaceae. Four families of seagrasses include some seventy species worldwide, all flowering plants with closer affinities to lilies than to the true grasses. Seagrasses provide important ecological services: they trap sediments; stabilize muddy substrates; lessen the flow of storm-driven water; serve as cover and nursery areas for scores of species, including several of commercial importance (e.g., bay scallops); and provide food for green turtles, manatees, and waterfowl. Clearly, any misfortune befalling seagrasses also produces far-reaching effects that extend throughout an entire ecosystem.

produces enzymes that dissolve the plant's cells, first on the leaves, where dark spots develop and soon blacken the entire leaf blade before moving to the rootstocks. In light of such damage, biologists aptly dubbed the malady as eelgrass wasting disease, and some, with only a bit of exaggeration, characterized the event as the "botanical Black Death," a clear reference to the plague that devastated medieval Europe. Indeed, the plants, once forming dense underwater meadows, died en masse, leaving behind only remnants at locations where salinities were low enough to become a barrier to further spread of the disease. In the process, populations of species dependent on eelgrass plummeted, not the least of which were bay scallops and Atlantic brant. Another species, a limpet, silently vanished into the cemetery of extinction. To assess the impact on brant, consider that eelgrass formed 85 percent of their diet prior to 1932 but just 9 percent in the following years. Eelgrass eventually recovered, but, as described below, the submersed vegetation in North Carolina's inland seas still faces significant threats.

Unfortunately, runoff flowing into the rivers and later into the sounds carries with it sediments that can smother the plants (and oyster reefs) and, of even greater importance, excessive amounts of nutrients originating from fertilizers and the organic wastes from residential and industrial sources. These stimulate thick algal blooms that diminish the amount of light available to the plants, at times creating lifeless wastelands where benthic communities once flourished. Severe algal blooms often develop where the Chowan and Neuse Rivers flow, respectively, into Albemarle and Pamlico Sounds. Efforts to curtail the nutrient-rich runoff remain at the forefront of conserving the submersed vegetation and the recovery of benthic life it supports. Unfortunately, hurricanes, an irregular but recurring factor along the coast, may increase runoff because of heavy rainfall and flooding, which then introduce even heavier sediment and nutrient loads into the sounds.

Currituck Sound, lush with pondweeds, widgeongrass, wild celery, and other freshwater vegetation in its upper reaches, offers an instructive example of anthropogenic and natural influences on submersed vegetation. Because of the latter, the sound became a major wintering area for waterfowl in the Atlantic Flyway and accordingly attracted legions of hunters, many of which were wealthy industrialists from northern states. The opulent Whalehead, a residence built by a husband-and-wife team of avid duck hunters, now serves as a county-owned tourist attraction and site for civic events.

The freshwater marshes of cattails and bulrushes that border Currituck Sound also provide important habitat for a viable population of king rails, which have declined precipitously elsewhere in their range. These wetlands, including those in the care of Mackay Island National Wildlife Refuge, may indeed be the best habitat for king rails anywhere along the Atlantic coast. A three-to-five-year cycle of controlled burning at the refuge helps improve the quality of the marshes for the rails and other wildlife (e.g., retarding the intrusion of woody vegetation).

Currituck Sound has not remained unaffected by dynamic influences, beginning with the expansion of the Albemarle and Chesapeake Canal between 1914 and 1918, later a segment of the Intracoastal Waterway. Salt water and pollution from Norfolk Harbor flowed directly into the northern end of the sound, which soon produced harmful effects, including a bottom covered by a deep layer of sludge, which accumulated by as much as two inches per year. By 1926, large areas were nearly devoid of any aquatic vegetation. In 1932, new locks at the north end of the canal, although designed primarily to keep salt water from entering the canal, also curtailed the inflow of sewage and industrial wastes. As a result, the vegetation in the sound recovered and regained its former abundance.

In 1962, a severe nor'easter opened several inlets and allowed seawater to enter the sound. These quickly closed, but two years later, fishermen reported a "strange plant" in Currituck Sound; by 1966, mats of Eurasian watermilfoil covered nearly 66,700 acres and had spread into adjacent parts of Back Bay and Albemarle Sound. The explosion of Eurasian watermilfoil seems related to the sudden increase in salinity, which, through the flocculation of suspended sediments, temporarily diminished turbidity in the water column. This favored the rapid growth of an exotic plant that could outcompete native vegetation adapted to natural levels of turbidity. In places, Eurasian watermilfoil still crowds out the native vegetation, although it does provide desirable food for some species of waterfowl.

North Carolina's Heritage of Duck Hunting and Decoys

North Carolina's sounds earned a revered place in the annals of waterfowl hunting, which peaked, give or take, between 1875 and 1935. In particular, wealthy industrialists from the Northeast formed hunting clubs and built lodges, which once lined the shores of Currituck Sound. Regrettably, the era also included market hunting, when, in the absence of federal oversight, countless thousands of waterfowl and other migratory birds were shot each year for the cuisine of fancy restaurants in eastern cities. Between 1903 and 1909, this practice put at least $100,000 each year into the pockets of Currituck's market hunters. In the winter of 1910–11 alone, nearly 250,000 birds were shipped from Currituck Sound to Norfolk, Virginia, and similar tallies occurred elsewhere (e.g., Chesapeake Bay). The slaughter, often enabled by large-bore punt guns, baiting, and other exploitative tactics, ended with enactment of the Migratory Bird Treaty Act in 1918. Not incidentally, decoys and decoy carvers emerged during this era and, in doing so, unknowingly established a uniquely American form of folk art. Today, avid collectors willingly pay thousands of dollars for decoys made by certain carvers, many of whom were watermen, trappers, and boat builders seeking additional income.

Native Americans were the first to employ decoys, as shown by the discovery in Lovelock Cave, Nevada, of eleven replicas of canvasbacks constructed of reeds and decorated with feathers; radiocarbon dating indicates they were crafted about 2,160 years ago by a now extinct Indian civilization known as Si-Te-Cah, or "tule eaters." In North Carolina, the tradition flourished in the work of carvers who fashioned sturdy and often oversized decoys designed to improve their visibility on the sounds' expansive waters (Fig. 4-11). Most were produced with a no-frills focus on function instead of artistic detail; some were indeed rather crude, whereas others represented elegant simplicity. Among the crudest are "root heads," whose heads and necks were fashioned from a single piece of naturally curved wood; these required little or no shaping and emphasized the rustic craftsmanship of "Down East" carvers. Others fabricated lightweight decoys with bodies of canvas-covered wire frames attached to a flat bottom board, a style used in the Outer Banks region more than anywhere else in North America. Goose and swan decoys made in this way were the hallmark of Tilman Lewark (1864–1945) of Corolla; another carver was Ned Burgess (1863–1956) of Churches Island, who also produced canvas-covered ducks. Among other differences in style, the number of wire "ribs" used to support the canvas body helps identify the work of these and other carvers. Nonetheless, to accommodate rough handling, most of the region's decoys were solid wood, not hollowed out, as was the custom at some locations (e.g., Barnegat Bay, New Jersey).

Artwork was minimal, as the harsh conditions on the sounds required frequent repainting of the decoys. Still, a few carvers skillfully painted important features, such as a bright wing patch—the speculum—of northern pintails and other puddle ducks. In particular, decoys carved by Mitchell Fulcher (1869–1950) featured exceptional artwork that today commands high prices from collectors. Fulcher, who lived at Stacy on the shores of Core Sound, sought white pine from the masts and spars of shipwrecks as stock for his decoys; this easy-to-work straight-grained wood was otherwise unavailable on the Coastal Plain. Because of his skills with a pocket knife, he also supplied heads for the decoys of other carvers. Using dead ducks as models, Fulcher outlined his patterns for heads on the lids of shoeboxes. Lesser scaup, locally known as blackheads, predominated his output of waterfowl decoys, but he (as did others) also carved dowitchers and other shorebirds.

Among many others, the work of twin brothers Lem and Lee Dudley (born 1861) at Knotts Island deserves mention, as does Alvirah Wright (1869–1951) at Duck and James Best (1866–1933) of Kitty Hawk. Like Fulcher, Best at times carved decoys from the masts of wrecked schooners, underscoring the region's mournful distinction as the Graveyard of the Atlantic.

Cast-iron "wing ducks" hold a special place in the history of decoys. These weighed-down "wings" were platforms that extended from the edges of a sinkbox, a waterproof, coffin-like structure that enclosed a gunner. As a result of the additional weight, the so-called battery sank nearly flush with the surface and, without much profile, essentially disappeared from view. Dozens, even hundreds, of conventional wooden decoys were deployed around the battery, which became an aquatic killing ground for unlimited numbers of ducks. As many as fifty batteries once operated in Currituck County and more than twenty more in Dare County. As advertised in its 1913 catalog, the Elizabeth City Iron Works (established in 1875) cast and sold unpainted wing decoys in twenty- and thirty-pound sizes, respectively priced at fifty and seventy-five cents, as well as "duck uprights" (stabilizing weights attached to the bottom of wooden decoys) for two and a half cents each. Some of the wing decoys at times fell overboard and are occasionally dredged up by oystermen.

FIGURE 4-11. Decoys produced in North Carolina are among those now prized by collectors. *Top, left to right,* a root-head brant by Stacy Howard (1885–1968) and a canvasback by Theodore Williams (1853–1926). *Center,* a pintail dating to about 1895 by famed carver Mitchell Fulcher (1869–1950). *Bottom, left to right,* a cast-iron wing decoy and a contemporary canvas-covered Canada goose. Photo credit: Robbie Smith (*center*).

Oysters: More Than Just Good Eating

Oysters, eaten raw or roasted and added to stews or casseroles, have long appealed to humans, as witnessed by middens—accumulations of empty shells piled ever higher by generations of Native Americans. Middens reveal the availability of various kinds of shellfish, which may in turn reflect changes occurring long ago in coastal environments. In places, middens also create local conditions where a distinctive flora may develop at the site. In other cases, the on-site vegetation becomes more robust because the shells enrich the soil with calcium and phosphorous.

More than just a good meal, oysters also provide ecological services, among them their remarkable capacity to filter water; one oyster can process as much as fifty gallons per day in search of food. In doing so, they clear water of microalgae, organic nutrients, and suspended particles, thereby improving light penetration in the water column and lessening the harmful effects of eutrophication (e.g., algal blooms). The same process, however, may concentrate toxic materials, heavy metals, or pathogenic bacteria in their tissues and jeopardize the health of humans eating oysters harvested from contaminated waters.

Oysters also provide sounds and estuaries with structures—beds or reefs—that benefit a diverse assortment of marine life that, in due course, eventually develops into a fully interactive community (Fig. 4-12). These organisms in turn attract predators, many of which are gamefish prized by anglers. Significantly, oyster beds also serve as a platform for the development of future generations of oysters (see below). A viable bed continues to expand not only laterally but also upward, thus becoming a three-dimensional structure of nooks and crannies ready-made for occupancy. This complex architecture adds about fifty times more surface area in comparison with a featureless benthic environment of the same size. Because the tops of some reefs may be partially exposed at low tide, they become feeding grounds for wading birds, of course including the aptly named American oystercatcher.

One of the common reef dwellers, the flatback mud crab, began dealing with one of nature's more hideous yet fascinating parasites that reached the Atlantic coast during the 1960s—a parasitic barnacle, *Loxothylacus*

FIGURE 4-12. Oyster beds provide habitat for a vibrant community of other species, protect shorelines from erosion, and maintain a natural substrate where new generations of oyster larvae mature year after year. Photo credit: Troy Alphin, UNCW Benthic Ecology Laboratory.

panopaei, or more simply "Loxo," that arrived from the Gulf of Mexico as a stowaway in a shipment of oysters. After attaching to their hosts, female Loxos turn mud crabs of either sex into castrated "zombies" by penetrating their internal reproductive organs and brains with root-like structures. They also produce an external reproductive sac called an externa on the abdomen of the infected crabs that mimics the appearance of the actual egg mass or "sponge" spawned and carried by healthy female crabs. Infected female mud crabs thus devote their care to the parasites' offspring instead of their own. Remarkably, so do infected males, which would otherwise not exhibit this form of maternal behavior. In short, Loxos hijack the crabs' life cycle. Infection rates range as high as 90 percent in some areas, and with population densities of up to about fifty mud crabs per square yard, an important member of the reef community may be in jeopardy. Birds, fishes, and blue crabs forage on mud crabs, which themselves prey on tiny bivalves and other crustaceans that thrive on healthy oyster reefs.

Few species can match the reproductive output of oysters. A single female may issue 5 million eggs and one male as many as 2.5 billion sperm—gametes that disperse randomly in the water column. As such, this strategy entails "flooding the market," in which fertilization relies on chance encounters of the drifting gametes. After hatching, the eggs produce the first of three larval stages, ending with those known as spats that settle permanently on hard surfaces. By nature's design, an existing reef ideally fulfills that need, although rocks and other hard surfaces may also serve as "cultch" for spats. Still, many larvae die simply because they land on unsuitable substrates, and only one spat in a million may find a suitable footing. Those that survive will reach the size of a dime in three months and grow into mature oysters in about three years.

Some oysters may be invaded by the pea crab, a kleptoparasite so named because it steals food, in this case by gleaning microorganisms captured on the gills of its host. Larval pea crabs of both sexes initially infect spats, where they mature in concert with their hosts, but after breeding, only females remain; males leave in search of mates elsewhere. Depending on location, pea crabs may infect as many as 73 percent of the oyster population. Pea crabs damage the gills of their hosts, not fatally, but enough to slow their growth, especially when food is not abundant. For some oyster-loving gourmets, the small pinkish crabs offer an added treat, notably including George Washington, who sprinkled pea crabs atop oyster soup. For the rest of us, oysters will do just fine "as is," thanks to some unknown ancestor who, according to Jonathan Swift in *Gulliver's Travels*, "was a brave man that first ate an oyster." Yet, there's a touch of irony here—removing an oyster to eat also removes its habitat.

Oysters are the focus of two conservation efforts that share a common modus operandi. The first concerns the restoration of North Carolina's depleted supply of oysters by establishing new reefs at key locations. To accomplish this goal, hundreds of tons of crushed concrete, granite, and limestone along with thousands of bushels of oyster shells dumped on the floor of sounds and estuaries become a starting point. These materials, especially the limestone, a source of calcium carbonate for mollusk shells, furnish the cultch for oyster larvae, which, when mature, form a new reef. Some of these reefs will eventually be open to regulated harvests, whereas others will remain sanctuaries dedicated solely to larval production (e.g., the recently completed forty-acre Swan Island Oyster Sanctuary in Pamlico Sound near the mouth of the Neuse River).

Second, "fringing" reefs, which are initiated with loose or bagged shells, stem the erosion of vulnerable shorelines. Volunteers secure mesh bags filled with empty clam or oyster shells in rows parallel to the shore, where, as cultch, they become the building blocks for a new reef. The result is a "living shoreline" that lessens the destructive power of storm-driven waves and rising sea levels. Hence the popular saying "if you shuck it, don't chuck it" advocates recycling shells for shoreline protection.

Meanwhile, efforts are under way to develop and support a thriving oyster mariculture industry with the potential of adding $100 million to the regional economy. To succeed, however, each of these projects, whether for environmental or economic objectives, depends on reducing the polluted stormwater runoff entering the sounds and estuaries that rim North Carolina's extensive coastline.

Sea Ducks: A Menagerie of Waterfowl

Pamlico and Albemarle Sounds provide winter habitat for several species of sea ducks, a diverse group that includes mergansers, eiders, goldeneyes, and scoters, as well as the long-tailed duck (once known as an old squaw), the harlequin duck, and the extinct (since 1875) Labrador duck. The bill structure differs greatly within the group, as shown by the tooth-like serrations of mergansers, which facilitate the capture of fishes; others have stout bills with robust tips, which enable removal of mussels attached to hard surfaces. Indeed, animal matter dominates the diet of all the sea ducks, and, to that end, all are accomplished divers. Long-tailed ducks, in fact, can dive to depths of 180 feet, an exploit that necessitates using their wings to propel them back to the surface (an ability lacking in any other species of waterfowl). Some species of sea ducks take two to three years to reach sexual maturity and live for many years. Despite the diversity within the group, taxonomists nonetheless unite species in a single tribe, a taxon between genus and family that has long served to distinguish certain groups of waterfowl.

Three species of scoters overwinter in the sounds; of these, black and surf scoters occur most often, the white-winged scoter less frequently (Fig. 4-13). Female scoters are not readily separated by species under field conditions, whereas the males can be identified more easily. The dark bill of male black scoters features an enlarged yellow knob at its base, which differs from the orange bill with a black knob of male white-winged scoters. Male surf scoters outdo both; their multicolored—white, orange, and black—bills set them apart, as do patches of white on their crowns and napes (a feature leading some to call the birds "skunk-heads"). White-winged scoters (both sexes) of course sport their namesake markings, although these may not always be visible on swimming birds. Otherwise, all three have black body plumage.

Scoters dive in open water primarily for mollusks but also for small crustaceans and marine worms. At night, they rest well offshore in rafts that may include scores of birds. Because they frequent expanses of open water, sea ducks remain especially susceptible to oil spills. Scoters nest in freshwater habitats, often at the edges of ponds and lakes, in boreal

FIGURE 4-13. Black scoters, distinguished by prominent yellow knobs on their bills, are among the sea ducks wintering on North Carolina's sounds and inshore ocean waters.

forest and tundra zones. They all lay relatively small clutches of six to ten eggs and lack the adaptability of many other ducks (e.g., mallards) to cope with droughts or other changes in local environmental conditions. As with some other species of sea ducks, scoter broods may include ducklings produced by other females—a characteristic known as brood amalgamation. In the case of white-winged scoters, for example, one or two females may care for broods with as many as 55 ducklings; amazingly, one study reported that up to seven females attended broods as large as 150 ducklings. More often, however, amalgamated broods number fifteen to twenty-five ducklings. For reasons still unclear, scoter populations have substantially decreased in recent decades.

Duck hunters normally take great pains to remain inconspicuous when luring their quarry into shooting range, not the least of which is to remain hidden and motionless. Not so when hunting scoters. Instead, hunters wave black flags! Scoters en route to another location, on seeing a waving flag, will change course and head for the blind (a camouflaged boat anchored well offshore). How hunters discovered this strategy remains unknown, but in behavioral terms, it relies on what is known as an approach response, which brings an organism closer to a stimulus—in this case, the visual simulation of the wing movements of birds about to land. The flags work because they provide a stimulus visible to low-flying scoters across a wide sweep of open water. As the birds approach the blind, a string of decoys completes the deception.

Hardly Mermaids

It takes a lot of imagination to envision a plump ten-to-thirteen-foot sea creature with paddle-like forelimbs and a face covered with stiff bristles as an attractive woman, albeit one that is half fish. Nonetheless, early mariners, perhaps long overdue for shore leave, thought just that and thereby gave rise to fables of mermaids. Manatees, like whales, are totally aquatic mammals, but unlike whales, they evolved independently from land-based ancestors that also produced elephants.

Florida manatees, although semitropical, also visit North Carolina between June and October when water temperatures along the coast

Safe Haven: The Intracoastal Waterway

In 1808, Albert Gallatin (1761–1849), the Swiss-born secretary of the treasury in the cabinet of Thomas Jefferson, proposed construction of a protected waterway that would connect Massachusetts and Georgia, then the southernmost state in the new nation. The route relied on natural bodies of water for most of the way and required just four canals to cross intervening barriers. Hence, less than one hundred miles of canals could provide the nation with inland navigation along the greater part of its coast, protected against "storms and enemies" and useful in "war and peace." Nonetheless, Gallatin's far-sighted plan did not gain the political support necessary for its implementation (the War of 1812 likewise hampered its initiation), but his idea survived, albeit taking more than a century to reach fruition.

One of the canals in Gallatin's plan cut across the wetlands between Chesapeake Bay and Albemarle Sound. In fact, a canal initially surveyed by George Washington and completed in 1805 (to haul logs) already crossed the Great Dismal Swamp, although it was not in full use until 1829. In 1859, a better route through this region of swamps and marshes superseded the commercial usefulness of its narrow predecessor. Known as the Albemarle and Chesapeake Canal, it actually connects with Currituck Sound, but as the saying goes, "that's close enough for government work." Both remain in use today, but despite its history, only a few recreational boaters today select the Dismal Swamp Canal.

The current route of the Atlantic Intracoastal Waterway crosses the Albemarle Peninsula, thereby avoiding the risk of sailing the full length of Pamlico Sound in open water; this segment, in part, utilizes the Alligator River but required a canal to complete the crossing (Fig. 4-14). Other canals in North Carolina also shorten the route and avoid potentially hazardous travel. One of these connects the Neuse River Estuary with Morehead City, and another, Snow's Cut, provides a shortcut across the Cape Fear Peninsula, as well as bypassing the offshore perils of Frying Pan Shoals. The last segment of the waterway in North Carolina, which extends southward into South Carolina, was completed in 1936.

Unfortunately, canals sometimes produce unintended consequences, as occurred when sea lampreys gained access to the upper Great Lakes after the Welland Canal was enlarged in 1932. As a result, the eel-like parasites caused the collapse of a valuable lake trout fishery. Or when the Point Pleasant Canal, a segment of the Intracoastal Waterway completed in 1926, introduced salt water into the upper end of Barnegat Bay in New Jersey and quickly changed a freshwater "paradise" into a marine community. Similarly, the Albemarle and Chesapeake Canal allowed the influx of salt water into Currituck Sound, which, along with sewage and industrial wastes from Norfolk Harbor, killed much of the sound's submerged aquatic vegetation—a regrettable situation that ended in 1932 when new locks were installed at the canal's north end.

On the other hand, the massive amounts of dredged materials—"spoil"—that were removed and deposited during construction and maintenance of the channels formed islands that attracted large numbers of birds, especially colony-nesting gulls, terns, pelicans, herons, and egrets. Some species of shorebirds regularly nest on the upland sites, where the nesting densities of willets and American oystercatchers may be quite high. The interiors of some spoil islands, diked at their edges to retard erosion, become wetlands where other shorebirds stop and forage during their spring migrations. In the earliest stage of succession, the fresh deposits of spoil support little vegetation, but in time their once-barren surfaces progressively develop into grassy patches, shrubby thickets, and young forests, with each stage featuring its own community of birds. In all, 190 species of birds, or nearly half of all those known from the coastal zone of North Carolina, have been recorded on spoil islands. The Intracoastal Waterway also provides a migration route for manatees traveling between North Carolina and Florida, whereas others migrate along the coastline.

FIGURE 4-14. One of the more prominent land cuts on the Intracoastal Waterway links Albemarle and Pamlico Sounds (*top*). Elsewhere along the waterway in North Carolina, piles of material dredged from coastal marshes or the Cape Fear River form spoil islands that offer prime nesting habitat for several species of colonial waterbirds, including royal terns (*bottom*).

exceed sixty-eight degrees (i.e., typically between June and October). Not all manatees do so, but those that undertake the migration sometimes take advantage of the Intracoastal Waterway, whereas others follow the coastline. Sightings of manatees in North Carolina occur at several locations, including along the shores of the Outer Banks, along the mainland shores of sounds, in the Intracoastal Waterway, and in estuaries; they often visit marinas at these sites. A few manatees travel considerable distances inland, including a sighting nearly sixty-nine miles up the Cape Fear River and another almost the same distance up the Neuse River. By November, most have left North Carolina for their winter quarters in Florida, a response triggered by the onset of cooling water. Those that remain will likely die from cold stress—a mix of emaciation, skin and intestinal diseases, and diminished metabolic and immunological functions initiated when water temperatures fall below sixty-eight degrees. Of nine standings reported between 1991 and 2012, seven were recorded between November and January.

Manatees feed exclusively on submerged aquatic vegetation; eelgrass, widgeongrass, and shoalgrass make up a large part of their diet. Each day they may consume up to 20 percent of their body weight, often leaving behind a trail through the vegetation known as a feeding scar. They chew their food with cheek teeth that replace themselves in a conveyer belt–like fashion; older worn teeth move forward and drop out as new teeth from the rear move forward. Although large, manatees pose no threats to humans, but the reverse is all too true, notably from deep, sometimes fatal cuts from motorboat propellers. Because of their limited numbers—currently fewer than five thousand—Florida manatees appear on the federal list of endangered species.

The Roanoke River and Its Namesake Refuge

Anonymous. 2013. Roanoke River National Wildlife Refuge habitat management plan. Reference code 48752, U.S. Fish and Wildlife Service Regional Office, Atlanta, Ga.

Ashley, K. W., and R. T. Rachels. 1999. Food habits of bowfin in the Black and Lumber Rivers, North Carolina. Proceedings of the Annual Conference of Southeastern Fish and Wildlife Agencies 53:50–60.

Bevelander, G. 1934. The gills of *Amia calva* specialized for respiration in an oxygen deficient habitat. Copeia 1934:123–127.

Bolen, E. G., and D. Flores. 1993. The Mississippi Kite: Portrait of a Southern Hawk. University of Texas Press, Austin.

Brantley, C. G., and S. G. Platt. 2001. Canebrake conservation in the southeastern United States. Wildlife Society Bulletin 29:1175–1181.

Chartier, N. A. 2014. Breeding biology of Swainson's warbler (*Limnothlypis swainsonii*) in a North Carolina bottomland hardwood forest. PhD dissertation, North Carolina State University, Raleigh.

Hendrick, M. S., S. L. Katz, and D. R. Jones. 1994. Periodic air-breathing behavior in a primitive fish revealed by spectral analysis. Journal of Experimental Biology 197:429–436.

Rahn, H., K. B. Rahn, B. J. Howell, et al. 1971. Air breathing of the garfish (*Lepisosteus osseus*). Respiratory Physiology 11:285–307.

Sallabanks, R., J. R. Walters, and J. A. Collazo. 2000. Breeding bird abundance in bottomland hardwood forests: habitat, edge, and patch size differences. Condor 102:748–758.

Saracco, J. F., and J. A. Collazo. 1999. Predation on artificial nests along three edge types in a North Carolina bottomland hardwood forest. Wilson Bulletin 111:541–549.

Scharph, C. 2013. Bowfin: North America's freshwater thug. American Fishes, Fish and Aquatic Conservation, U.S. Fish and Wildlife Service, Washington, D.C.

Wharton, C. H., W. M. Kitchens, E. C. Pendleton, and T. W. Sipe. 1982. The ecology of bottomland hardwood swamps of the southeast: a community profile. FWS/OBS-81-37, Biological Services Program, U.S. Fish and Wildlife Service, Washington, D.C.

Dodson, J. J., and W. C. Leggett. 1974. Role of olfaction and vision in the behavior of American shad (*Alosa sapidissima*) homing to the Connecticut River from Long Island Sound. Journal of the Fisheries Research Board of Canada 31:1607–1619.

Gerard, P. 1999. The Cape Fear: historic gateway to the Atlantic. Wildlife in North Carolina 63(11):82–91. [A portrait of the state's longest river.]

Gilbert, C. R. 1989. Species profiles: life histories and environmental requirements of coastal fisheries (Mid-Atlantic)—Atlantic and shortnose sturgeons. U.S. Fish and Wildlife Service Biological Report 82 (11.122), Washington, D.C.

Kynard, B., S. Bolden, M. Kieffer, et al. 2016. Life history and status of shortnose sturgeon (*Acipenser brevirostrum*, LeSueur, 1818). Journal of Applied Fisheries 32:208–248.

Leggett, W. C., and R. R. Whitney. 1972. Water temperature and the migrations of American shad. Fisheries Bulletin 70:659–670.

MacKenzie, C., L. S. Weiss-Glanz, and J. R. Moring. 1985. Species profiles: life histories and environmental requirements of coastal fishes and invertebrates (mid-Atlantic)—American shad. U.S. Fish and Wildlife Service Biological Report 82 (11.37), Washington, D.C.

McPhee, J. 2002. The Founding Fish. Farrar, Straus and Giroux, New York. [Describes the role of shad in Washington's choice of Valley Forge.]

Moser, M. L., and S. W. Ross. Habitat use and movements of shortnose and Atlantic sturgeons in the lower Cape Fear River, North Carolina. Transactions of the American Fisheries Society 124:225–234.

Neves, R. J., and L. Depres. 1979. The oceanic migration of American shad, *Alosa sapidissima*, along the Atlantic coast. Fisheries Bulletin 70:659–670.

Raabe, J. K., J. E. Hightower, T. E. Ellis, and J. J. Facendola. 2019. Evaluation of fish passage at a nature-like rock ramp fishway on a large coastal river. Transactions of the American Fisheries Society 148:798–816.

Rachels, K. T., and C. W. Morgeson. 2018. Cape Fear River striped bass spawning stock survey and proposed establishment of put-grow-take fishery. Federal Aid in Sport Fishery Project Report F-108, North Carolina Wildlife Resources Commission, Raleigh.

Smith, J. A., and J. E. Hightower. 2012. Effect of low-head lock-and-dam structures on migration and spawning of American shad and striped bass in the

Cape Fear River, North Carolina. Transactions of the American Fisheries Society 141:402–413.

Sorenson, C. 2019. How do sturgeon find their prey? Wildlife in North Carolina 83(6):43.

Sulak, K. J., R. E. Edwards, G. W. Hill, and M. T. Randall. 2002. Why do sturgeon jump? Insights from acoustic investigations of the Gulf sturgeon in the Suwannee River, Florida, USA. Journal of Applied Ichthyology 18:617–620.

Blackwater Streams: A Coastal Specialty

Benke, A. C., R. L. Henry III, D. M. Gillespie, and R. J. Hunter. 1985. Importance of the snag habitat for animal production in a southeastern stream. Fisheries 10:8–13.

Brownlow, C. A., and E. G. Bolen. 1994. Fish and macroinvertebrate diversity in first-order blackwater and alluvial streams in North Carolina. Journal of Freshwater Ecology 9:261–270.

Meffe, G. K., and A. L. Sheldon. 1988. The influence of habitat structure on fish assemblage composition in southeastern blackwater streams. American Midland Naturalist 120:225–240.

Meyer, J. L. 1990. A blackwater perspective on riverine ecosystems. BioScience 40:643–651.

Smock, L. A., and E. Gilinsky. 1992. Coastal plain blackwater streams. Pages 271–311 in Biodiversity of the Southeastern United States: Aquatic Communities (C. T. Hackney, S. M. Adams, and W. H. Martin, eds.). John Wiley & Sons, New York.

Smock, L. A., E. Gilinsky, and D. L. Stoneburner. 1985. Macroinvertebrate production in a southeastern U.S. blackwater stream. Ecology 66:1491–1503.

Ancient Trees on the Black River

Barbour, R. W., and W. H. Davis. 1969. Bats of America. University Press of Kentucky, Lexington. [Describes the relationship between Seminole bats and Spanish moss.]

Bat Conservation International and Southeastern Diversity Network. 2013. A conservation strategy for Rafinesque's big-eared bat (*Corynorhinus rafinesquii*) and southeastern myotis (*Myotis austroriparius*). Bat Conservation International, Austin, Tex. [A primary source for the biology, ecology, and management of the title species.]

Boyles, J. G., P. M. Cryan, G. F. McCracken, and T. H. Kunz. 2011. Economic importance of bats to agriculture. Science 332:451–452.

Faris, J. 1994. Old forests may be last refuge for rare bat. Coastwatch (March/April):16–21.

Fitzpatrick, J. W., M. Lammertink, M. D. Luneau Jr., et al. 2005. Ivory-billed woodpecker (*Campephilus principalis*) persists in continental North America. Science 308:1460–1462. [Several authorities refute the evidence leading to this conclusion.]

Frick, W. F., J. F. Pollock, A. C. Hicks, et al. 2010. An emerging disease causes regional population collapse of a common North American bat species. Science 329:679–682.

Loeb, S. C., M. L. Lacki, and D. A. Miller, eds. 2011. Conservation and management of eastern big-eared bats—a symposium. General Technical Report SRS-145, USDA Forest Service, Southern Research Station, Asheville, N.C.

Stahle, D. W., M. K. Cleaveland, D. B. Blanton, et al. 1998. The Lost Colony and Jamestown droughts. Science 280:564–567.

Stahle, D. W., M. K. Cleaveland, R. D. Griffin, et al. 2006. Decadal drought effects on endangered woodpecker habitat. Eos 87:121–125.

Stahle, D. W., M. K. Cleaveland, and J. G. Hehr. 1988. North Carolina climate changes reconstructed from tree rings: AD 372 to 1985. Science 240:1517–1519.

Stahle, D. W., J. R. Edmondson, I. M. Howard, et al. 2019. Longevity, climate sensitivity, and conservation status of wetland trees at Black River, North Carolina. Environmental Research Communications 1 041002.

Warren, L. 2004. Constantine Samuel Rafinesque: A Voice in the American Wilderness. University Press of Kentucky, Lexington.

Whitaker, J. O., Jr., and C. Ruckdeschel. 2010. Spanish moss, the unfinished chigger story. Southeastern Naturalist 9:85–94.

Young, O. P., and T. C. Lockley. 1989. Spiders of Spanish moss in the delta of the Mississippi. Journal of Arachnology 17:143–148.

Another Lost Colony?

Allen-Grimes, A. W. 1982. Breeding biology of the white ibis (*Eudocimus albus*) at Battery Island, North Carolina. MS thesis, University of North Carolina at Wilmington.

Bildstein, K. L. 1993. White ibis: Wetland Wanderer. Smithsonian Institution Press, Washington, D.C.

Frederick, P. C., and J. C. Ogden. 1997. Philopatry and nomadism: contrasting long-term movement behavior and population dynamics in white ibises and wood storks. Colonial Waterbirds 20:316–323.

Golder, W. W. 1990. Foraging flight activity of adult white ibis (*Eudocimus albus*) at a southeastern North Carolina colony. MS thesis, University of North Carolina at Wilmington.

Johnson, J. W., and K. L. Bildstein. 1990. Dietary salt as a physiological constraint in white ibis breeding in an estuary. Physiological Zoology 63:190–207.

Kushlan, J. A. 1979. Feeding ecology and prey selection in the white ibis. Condor 81:376–389.

Shields, M. A., and J. F. Parnell. 1983. Expansion of white ibis nesting in North Carolina. Chat 47:101–103.

Eels: A Slippery Travel Story

Anonymous. 2012. The American eel in North Carolina. Fact Sheet, Ecological Services Field Office, U.S. Fish and Wildlife Service, Raleigh, N.C.

Helfman, G. S., D. E. Facey, L. S. Hales Jr., and E. L. Bozeman Jr. 1987. Reproductive ecology of the American eel. American Fisheries Society Symposium 1:42–556.

Kruger, W. H., and K. Oliveira. 1999. Evidence for environmental sex determination of the American eel. Biology of Fishes 55:381–389.

Moser, M. L., W. S. Patrick, and J. U. Crutchfield Jr. 2001. Infection of American eels, *Anguilla rostrata*, by an introduced nematode parasite, *Anguillicola crassus*, in North Carolina. Copeia 2001:848–853.

Palstra, A. P., D. F. M. Heppener, V. J. T. van Ginneken, et al. 2007. Swimming performance of silver eels is severely impaired by swim-bladder parasite *Anguillicola crassus*. Journal of Experimental Marine Biology and Ecology 352:244–256.

Shepard, S. L. 2015. American eel. Biological Species Report FWS-HQ-ES-2015-0143, U.S. Fish and Wildlife Service, Hadley, Mass.

A Snail's Slide toward Extinction?

Adams, W. F. 1990. Magnificent rams-horn (*Planorbella magnifica*). Pages 27–30 *in* A Report on the Conservation Status of North Carolina's Freshwater and Terrestrial Mulluscan Fauna (W. F. Adams, ed.). Scientific Council on Freshwater Mollusks, North Carolina Wildlife Resources Commission, Raleigh.

Adams, W. F., and A. B. Gerberich. 1988. Rediscovery of *Planorbella magnifica* (Pilsbry) in southeastern North Carolina. Nautilus 102:125–126.

Bartsch, P. 1908. Notes on the fresh-water mollusk (*Planorbis magnificus*) and descriptions of two new forms of the same genus from the southern states. Proceedings of the United States National Museum 33:697–700.

Dall, W. H. 1907. Notes—*Planorbis magnificus*. Nautilus 22:90.

Lieb, B., K. Dimitrova, H.-S. Kang, et al. 2006. Red blood with blue blood ancestry: intriguing structure of a snail hemoglobin. Proceedings of the National Academy of Sciences 103:12011–12016.

Morgan, J. A. T., R. J. DeJong, Y. Jung, et al. 2002. A phylogeny of planorbid snails, with implications for the evolutions of *Schistosoma* parasites. Molecular Phylogenetics and Evolution 25:477–488.

Pilsbry, H. A. 1903. The greatest American *Planorbis*. Nautilus 17:75–76.

The Coastal Plain's Own Salamander

Ashton, R. E., Jr. 1985. Field and laboratory observations on microhabitat selection, movements, and home range of *Necturus lewisi* (Brimley). Brimleyana 10:83–106.

Ashton, R. E., Jr., A. L. Braswell, and S. I. Guttman. 1980. Electrophoretic analysis of three species of *Necturus* (Amphibia: Proteidae), and the taxonomic status of *Necturus lewisi* (Brimley). Brimleyana 4:43–46.

Brandon, R. A., and J. E. Huheey. 1985. Salamander skin toxins, with special reference to *Necturus lewisi*. Brimleyana 10:75–82.

Braswell, A. L., and R. E. Ashton Jr. 1985. Distribution, ecology, and feeding habits of *Necturus lewisi*. Brimleyana 10:13–35.

Cooper, J. E., and R. E. Ashton Jr. 1985. The *Necturus lewisi* study: introduction, selected literature review, and comments on the hydrologic units and their faunas. Brimleyana 10:1–12.

Sounds: North Carolina's Coastal Seas

Bangley, C. W., L. Paramore, D. S. Shiffman, and R. A. Rulifson. 2018. Increased abundance and nursery habitat use of the bull shark (*Carcharhinus leucas*) in response to a changing environment in a warm-water estuary. Scientific Reports 8, Article 6018, https://doi.org/10.1038/s41598-018-24510-z.

Carlton, J. T., G. J. Vermeij, D. R. Lindberg, et al. 1991. The first historical extinction of a marine invertebrate in an ocean basin: the demise of the eelgrass limpet *Lottia alveus*. Biological Bulletin 180:72–80.

Copeland, B. J., R. G. Hodson, and S. R. Riggs. 1984. The ecology of Pamlico River Estuary, North Carolina. FWS/OBS-82-06, Division of Biological Services, U.S. Fish and Wildlife Service, Washington, D.C.

Copeland, B. J., R. G. Hudson, S. R. Riggs, and J. E. Easley Jr. 1983. The ecology of Albemarle Sound, North Carolina: an estuarine profile. FWS/OBS-83-01, Division of Biological Services, U.S. Fish and Wildlife Service, Washington, D.C.

Cottam, C., J. J. Lynch, and A. L. Nelson. 1944. Food habits and management of American sea brant. Journal of Wildlife Management 8:36–56. [Highlights the crash of brant populations following the dramatic decline of eelgrass in the early 1930s. See Kirby and Obrecht (1982) for an alternate view of the brant population prior to the eelgrass die-off.]

Cottam, C. 1934. Eelgrass disappearance has serious effects on waterfowl and industry. Pages 191–193 *in* Yearbook of Agriculture. U.S. Department of Agriculture, Washington, D.C.

Davis, G. J., and M. K. Brinson. 1983. Trends in submersed macrophyte communities of the Carrituck Sound: 1909–1979. Journal of Aquatic Plant Management 21:83–87.

Davis, G. J., and D. F. Carey Jr. 1981. Trends in submersed macrophyte communities of the Currituck Sound: 1977–1979. Journal of Aquatic Plant Management 19:3–8.

Florschutz, O., Jr. 1972. The importance of Eurasian milfoil (*Myriophyllum spicatum*) as a waterfowl food. Proceedings of the Annual Conference of the Southeastern Association of Game and Fish Commissioners 26:189–193.

Fonesca, M. S., J. J. Fisher, J. C. Zieman, and G. W. Thayer. 1982. Influence of the seagrass, *Zostera marina* L., on current flow. Estuarine and Coastal Shelf Science 15:351–364.

Kirby, R. E., and H. H. Obrecht III. 1982. Recent changes in the North American distribution and abundance of wintering Atlantic brant. Journal of Field Ornithology 53:333–341.

Milne, L. J., and M. J. Milne. 1951. The eelgrass catastrophe. Scientific American 184(1):52–55.

Paerl, H. W., L. M. Valdes, A. R. Joyner, et al. 2006. Ecological responses to hurricane events in the Pamlico Sound system, North Carolina, and implications for assessment and management in a regime of increased frequency. Estuary and Coasts 29:1033–1045.

Quay, T. L., and T. S. Critcher. 1962. Food habits of waterfowl in Currituck Sound, North Carolina. Proceedings of the Annual Conference of the Southeastern Association of Game and Fish Commissioners 16:200–209.

Rogers, S. L., J. A. Collazo, and C. A. Drew. 2013. King rail (*Rallus elegans*) occupancy and abundance in fire managed coastal marshes in North Carolina and Virginia. Waterbirds 36:179–188.

Sand-Jensen, K. 1977. Effects of epiphytes on eelgrass photosynthesis. Aquatic Botany 3:55–63.

Short, F. T., L. K. Muehlstein, and D. Porter. 1987. Eelgrass wasting disease: cause and recurrence of a marine epidemic. Biological Bulletin 173:557–562.

Stauffer, R. C. 1937. Changes in the invertebrate community of a lagoon after disappearance of the eelgrass. Ecology 18:437–431.

Thayer, G. W., W. J. Kenworthy, and M. S. Fonseca. 1984. The ecology of eelgrass meadows of the Atlantic Coast: a community profile. FWS/OBS-84-02, U.S. Fish and Wildlife Service, Washington, D.C.

Wicker, A. M., and K. M. Endres. 1995. Relationship between waterfowl and American coot abundance with submersed aquatic vegetation in Currituck Sound, North Carolina. Estuaries 18:428–431.

Oysters: More Than Just Good Eating

Bahr, L. M., and W. P. Lanier. 1981. The ecology of intertidal oyster reefs of the south Atlantic coast: a community profile. FWS/OBS-81-15, Office of Biological Services, U.S. Fish and Wildlife Service, Washington, D.C.

Breitburg, D. L., L. D. Coen, M. W. Luckenbach, et al. 2000. Oyster reef restoration: convergence of harvest and conservation strategies. Journal of Shellfish Research 19:371–377.

Burrell, V. G. 1986. Species profiles: life histories and environmental requirements of coastal fishes and invertebrates (South Atlantic)—American oyster. U.S. Fish and Wildlife Service Biological Report 82 (11.570), Washington, D.C.

Byers, J. E., T. L. Rogers, J. H. Grabowski, et al. 2014. Host and parasite recruitment correlated at regional scale. Oecologia 174:731–738. [Infection rates of oysters by pea crabs at two sites in North Carolina varied from about 35 to 65 percent.]

Christensen, A. M., and J. J. McDermott. 1958. Live history and biology of the oyster crab, *Pinnotheres ostreum* Say. Biological Bulletin 114:146–179.

Claassen, C. 2017. Shell and shell-bearing sites in the Carolinas: some observations. Archaeology of Eastern North America 45:33–44.

Coen, L. D., R. D. Brumbaugh, D. Bushek, et al. 2007. Ecosystem services related to oyster restoration. Marine Progress Series 341:303–307.

Cook-Patton, S. C., D. Weller, T. C. Rick, and J. D. Parker. 2014. Ancient experiments: forest biodiversity and soil nutrients enhanced by Native American middens. Landscape Ecology 29:979–987.

Cressman, K. A., M. H. Posey, M. A. Mallin, et al. 2003. Effects of oyster reefs on water quality in a tidal creek estuary. Journal of Shellfish Research 22:753–762.

Davidson, A. 1979. North Atlantic Seafood. Viking Press, New York. [See page 206 regarding Washington's taste for pea crabs.]

Grabowski, J. H., R. D. Brumbaugh, R. F. Conrad, et al. 2012. Economic valuation of ecosystem services provided by oyster reefs. BioScience 62:900–909.

Hines, A. H., F. Alvarez, and S. A. Reed. 1997. Introduced and native populations of a marine parasitic castrator: variation in prevalence of the rhizocephalan *Loxothylacus panopaei* in xanthid crabs. Bulletin of Marine Science 61:197–214.

McAvoy, W. A., and J. W. Harrison. 2012. Plant community classification and the flora of Native American shell-middens on the Delmarva Peninsula. Maryland Naturalist 52:1–34.

Meyer, D. 1994. Habitat partitioning between the xanthid crabs *Panopeus herbstii* and *Eurypanopeus depressus* on intertidal oyster reefs (*Crassostrea virginica*) in southeastern North Carolina. Estuaries 17:674–679.

O'Shaughnessy, K. A., J. M. Harding, and E. J. Burge. 2014. Ecological effects of the invasive parasite *Loxothylacus panopaei* on the flatback mud crab *Eurypanopeus depressus* with implications for estuarine communities. Bulletin of Marine Science 90:611–621.

Trant, A. J., W. Nijland, K. M. Hoffman, et al. 2016. Intertidal resource use over millennia enhances forest productivity. Nature Communications 7, Article 12491, https://doi.org/10.1038/ncomms12491.

Sea Ducks: A Menagerie of Waterfowl

Baldassarre, G. 2014. Ducks, Geese, and Swans of North America. 2 vols. Johns Hopkins University Press, Baltimore, Md. [See volume 2 for life histories of scoters and other sea ducks.]

Brown, P. W., and M. A. Brown. 1981. Nesting biology of the white-winged scoter. Journal of Wildlife Management 45:38–45.

Goudie, R. I., M. R. Petersen, and G. J. Robertson, eds. 1999. Behavior and ecology of sea ducks. Canadian Wildlife Service Occasional Papers Series No. 100, Ottawa.

Kehoe, F. P., P. W. Brown, and C. S. Houston. 1989. Survival and longevity of white-winged scoters nesting in central Saskatchewan. Journal of Field Ornithology 60:133–136.

Krementz, D. G., P. W. Brown, F. P. Kehoe, and C. S. Houston. 1997. Population dynamics of white-winged scoters. Journal of Wildlife Management 61:222–227.

Livezey, B. C. 1995. Phylogeny and evolutionary ecology of modern sea ducks (Anatidae: Mergini). Condor 97:233–255.

McGilvrey, F. B. 1967. Food habits of sea ducks from the north-eastern United States. Wildfowl 18:141–145.

Savard, J.-P. L., and P. Lamothe. 1991. Distribution, abundance and aspects of the breeding ecology of black scoters, *Melanitta nigra*, and surf scoters, *Melanitta perspicillata*, in northern Quebec. Canadian Field Naturalist 105:488–496.

Silverman, E. D., D. T. Saafeld, J. B. Leiness, and M. D. Knoeff. 2013. Wintering sea duck distribution along the Atlantic coast of the United States. Journal of Fish and Wildlife Management 4:178–198.

Snell, R. R. 1985. Underwater flight of long-tailed duck (Oldsquaw) *Clangula hyemalis*. Ibis 127:267.

Hardly Mermaids

Cummings, E. W., D. A. Pabst, J. E. Blum, et al. 2014. Spatial and temporal patterns of habitat use and mortality of the Florida manatee (*Trichechus manatus latirostris*) in the Mid-Atlantic states of North Carolina and Virginia from 1991 to 2012. Aquatic Mammals 40:126–138.

Reep, R. L., and R. K. Bonde. 2006. The Florida manatee: biology and conservation. University Press of Florida, Gainesville.

Schwartz, F. J. 1995. Florida manatees, *Trichechus manatus* (Sirenia: Trichechidae), in North Carolina. Brimleyana 22:53–60.

Webster, W. D., J. F. Parnell, and W. C. Biggs Jr. 1985. Mammals of the Carolinas, Virginia, and Maryland. University of North Carolina Press, Chapel Hill.

Infobox 4-1: River-Bottom Parrots Forever Gone

Anonymous. 2009. Announcement of a false discovery of the extinct Carolina parakeet a mistake. http:/wildbirdsbroadcasting.blogspot.com. [Reveals the hoax of a population surviving in Honduras.]

Brewer, T. M. 1840. Wilson's American Ornithology. Otis, Broaders & Company, Boston. [A reprint published in 1970 by Arno Press, New York; see page 249 for Wilson's experiment with cats.]

Burgio, K. R., C. J. Carlson, and M. W. Tingley. 2017. Lazarus ecology: recovering the distribution and migration patterns of the extinct Carolina parakeet. Ecology and Evolution 7:5467–5475. [Offers a sharply defined range map based on historical records.]

Catesby, M. 1974. The natural history of Carolina, Florida, and the Bahama Islands. Beehive Press, Savannah, Ga. [Reprint of the third edition of Catesby's two-volume work, which was published, respectively, in 1732 and 1743.]

Cokinos, C. 2000. Hope Is a Thing with Feathers. Tarcher/Putnam, New York.

Greenberg, J. 2014. A Feathered River across the Sky: The Passenger Pigeon's Flight to Extinction. Bloomsbury, New York. [Vividly recounts the result of wanton shooting: extinction.]

Hahn, P. 1963. Where Is That Vanished Bird? An Index to the Known Specimens of the Extinct and Near Extinct North American Species. Royal Ontario Museum, University of Toronto Press, Toronto, Canada.

Laycock, G. 1969. The last parakeet. Audubon 71(March):21–25.

McKinley, D. 1979. Historical review of the Carolina parakeet in the Carolinas. Brimleyana 1:81–98.

Saikku, M. 1990. The extinction of the Carolina parakeet. Environmental History Review 14:1–18.

Snyder, N. F. 2004. The Carolina Parakeet: Glimpses of a Vanished Bird. Princeton University Press, Princeton, N.J.

Snyder, N. F., and K. Russell. 2002. Carolina Parakeet (*Conuropsis carolinensis*). The Birds of North America, No. 667 (A. Poole and F. Gill, eds.). The Birds of North America, Philadelphia, Pa.

Infobox 4-2: Red Tide Pays a Visit

Flewelling, L. J., J. P. Naar, J. P. Abbott, et al. 2005. Red tides and marine mammal mortalities. Nature 435:755–756. [Describes the lethal effects of toxins entering food chains.]

Magana, H. A., C. Contreras, and T. A. Villareal. 2003. A historical assessment of *Karenia brevis* in the western Gulf of Mexico. Harmful Algae
2:163–171.

Pietrafesa, L. J., G. S. Janowitz, K. S. Brown, et al. 1988. The invasion of the red
tide in North Carolina coastal waters. North Carolina Sea Grant Working
Paper 88-1, North Carolina State University, Raleigh.

Tester, P. A., R. P. Stumpf, F. M. Vukovich, et al. 1991. An expatriate red
tide bloom: transport, distribution, and persistence. Limnology and
Oceanography 36:1053–1061. [Describes the occurrence of red tide in North
Carolina.]

Walsh, J. J., J. K. Joliff, B. P. Darrow, et al. 2010. Red tides in the Gulf of
Mexico: where, when, and why? Journal of Geophysical Research 111:1–46.

Infobox 4-3: North Carolina's Heritage of Duck Hunting and Decoys

Andresen, K., and P. Harvey. 2015. Shorebird Decoys of North Carolina. Spark
Publications, Charlotte, N.C.

Day, A. M. 1959. North American Waterfowl. 2nd ed. Stackpole, Harrisburg, Pa.

Dudley, J., and K. Andresen, eds. 2011. Mitchell Fulcher, Master Decoy Carver.
Carolina Decoy Association, Cary, N.C.

Johnson, A. 1994. Canvas Decoys of North America. CurBac Press, Virginia
Beach, Va.

Johnson, A., and B. Coppedge. 1991. Gun Clubs and Decoys of Back Bay and
Currituck Sound. CurBac Press, Virginia Beach, Va.

Loud, L. L., and M. R. Harrington. 1929. Lovelock Cave. University of
California Publications in American Archaeology and Ethnology 25:1–183.

Nickens, T. E. 1999. Old Currituck. Coastwatch. Spring issue:6–13. North
Carolina Sea Grant, North Carolina State University, Raleigh.

North, D. C., Jr. 1990. North Carolina. Pages 168–183 *in* The Great Book of
Wildfowl Decoys (J. Engers, ed.). Thunder Bay Press, San Diego, Calif.

Tuohy, D. R., and L. Kyle Napton. 1986. Duck decoys from Lovelock Cave,
Nevada, dated by ^{14}C accelerator mass spectrometry. American Antiquity
51:813–816.

Infobox 4-4: Safe Haven: The Intracoastal Waterway

Bence, J. R., R. A. Bergstedt, G. C. Christie, et al. 2003. Sea lamprey
(*Petromyzon marinus*) parasite-host interactions in the Great Lakes.
Journal of Great Lakes Research 29:253–282.

Cammen, L. M., E. D. Seneca, and B. J. Copeland. 1974. Animal colonization of salt marshes artificially established on dredge spoil. North Carolina Sea Grant Publication UNC-SG-74-15, North Carolina State University, Raleigh.

Davis, G. J., and M. M. Brinson. 1983. Trends in submersed macrophyte communities of the Currituck Sound: 1909–1979. [Reports the harmful effects of the Albemarle and Chesapeake Canal on Currituck Sound.]

Goodrich, C. 1958. The Gallatin Plan after one-hundred fifty years. Proceedings of the American Philosophical Society 102:436–441.

Hinshaw, C. R., Jr. 1948. North Carolina canals before 1860. North Carolina Historical Review 25:1–56.

Parkman, A. 1983. History of the waterways of the Atlantic coast of the United States. Navigation History NWS-83-10, Institute for Water Resources, U.S. Army Corps of Engineers, Alexandria, Va.

Parnell, J. F., R. N. Needham, R. F. Soots Jr., et al. 1986. Use of dredged-material deposition sites by birds in coastal North Carolina, USA. Colonial Waterbirds 9:210–217.

Soots, R. F., Jr., and J. F. Parnell. 1975. Ecological succession of breeding birds in relation to plant succession on dredge islands in North Carolina estuaries. North Carolina Sea Grant Publication UNC-SG-75-27, North Carolina State University, Raleigh.

Van Deventer, F. 1964. Cruising New Jersey Tidewater: A Boating and Touring Guide. Rutgers University Press, New Brunswick, N.J. [Describes the freshwater "paradise" that was destroyed by a canal connecting a saltwater inlet with the upper end of Barnegat Bay.]

V

Interior Wetlands

"A Vast Body of Mire and Nastiness"

With those words, Colonel William Byrd II summed up his opinion of the Great Dismal Swamp, where he led a survey in 1728 to establish the boundary between Virginia and North Carolina. George Washington, however, thought it a "glorious paradise" when he visited the site in 1763. Divergent views, to be sure, but naturalists surely would side with the future president. Whereas the swamp straddles the state line, it represents a single ecological entity and is discussed here without further distinction of geopolitical locations.

The abrupt rise of the Suffolk Scarp, an ancient beachline, forms the western border of the Dismal Swamp, but to the east the land gently slopes until it reaches the coast. Since 1805, however, the Dismal Swamp Canal has cut off the eastward flow of the sluggish water so that a large part of the original swamp no longer exists; in all, likely only a third of the one million or more acres of wetland remain today. An impervious layer of clay seals the downward flow of water. Deep peat deposits, some up to thirteen feet thick, overlie the clay floor. In geological terms, Dismal Swamp is young: it began forming at the end of the Pleistocene some 12,000 to 11,000 years ago.

A community known as nonriverine swamp covers much of the area, so named because it depends on an accumulation of rainfall instead of an association with streams. Bald cypress, tupelo, and Atlantic cedar originally dominated the community, but years of logging have lessened their prevalence, which was further affected by ditching and fire suppression. Red maple, an extraordinarily adaptable species, filled the void and continues to expand. Extensive pocosins occur elsewhere in the swamp (more about pocosins follows).

Peat deposits underlie most of the forest vegetation and in the past fueled fires that maintained these communities. For example, fires kill Atlantic white cedar but also removes the upper layer of peat and enables the seeds of cedar to germinate on the exposed soil. (Ecologists refer to these residual accumulations as seed banks.) The result is a new generation of white cedars, often in pure stands. However, with fire suppression and lower water tables resulting from an extensive ditch system, the natural burning regime no longer checks red maple and other species (e.g., sweetgum) from increasing their presence.

Lake Drummond, named for its supposed discoverer (in 1655) and the first governor of the Albemarle Sound Colony, lies near the center of the remaining swamp. Oval in shape, the 3,140-acre freshwater lake also lies at the center of numerous theories regarding its origin. One suggests that an extensive peat fire hollowed out a depression that filled with water, whereas another posits that its basin formed from the impact of a meteor; if the latter, the lake may represent another Carolina bay (see below). Other theories include tectonic and mythical origins, but all remain controversial. Based on pollen in its bottom sediments, the lake is no more than four thousand years old and thus much younger than the swamp itself. Additionally, remnants of charred materials in the bottom sediments argue that a peat fire formed the lake basin.

The prevalence of peat in the Dismal Swamp also accounts for the lake's tea-colored water and its acidity; high turbidity limits light penetration downward in the water column. Once thought to be twelve to fifteen feet deep, Lake Drummond today is half its original depth, no doubt from the inflow of organic litter transported by the ditch system.

Habitat in the lake varies little; hence, relatively few fishes find niches to occupy. The fauna includes some twenty-six native species and six others that were introduced. Moreover, aquatic vegetation is sparse, which limits the food chain and energy needed to support a larger fish population and likely explains why some individuals are stunted. Likewise, waterfowl find little food and use the lake primarily as a resting area.

The biota includes only a few rare species of either plants or animals. Least trillium, silky camellia, and southern twayblade grow on some of the drier sites but are otherwise uncommon in the swamp. Dismal Swamp is home for about thirty species of ferns, of which the log fern is noteworthy—so named because, when discovered in the late 1800s, it was growing on a decomposing log. The species is rare and localized in eastern North America but occurs far more commonly in the Dismal Swamp than elsewhere in its range. Log ferns hybridize readily with others of its genus, which has fascinated botanists concerned with the evolutionary history of the group. At least one small mammal, the Dismal Swamp short-tailed shrew, evolved as a subspecies closely associated with the site (Fig. 5-1). Another, the Dismal Swamp southeastern shrew, was originally thought to be endemic and therefore was listed as threatened. However, subsequent surveys revealed that the subspecies has a wider distribution, and it was accordingly delisted. Curiously, both subspecies

FIGURE 5-1. A subspecies of the northern short-tailed shrew occurs in and near the Great Dismal Swamp. The largest shrew in North America, the species produces toxic saliva that paralyzes earthworms, slugs, insects, mice, and other prey that at times may be stored for later consumption.

George Washington's Ditch

In 1763, thirty-one-year-old George Washington, freshly home from the French and Indian War, visited the Great Dismal Swamp. He and eleven others had paid $20,000 for forty thousand acres of the vast wetland in what seems like an early version of the story about naive customers buying land sight unseen in certain parts of Florida. Nonetheless, while Washington regarded the swamp as a "paradise," he and the other investors hoped to reap profits from timber sales and from converting its mucky wetlands into farmland in a venture incorporated as the Dismal Swamp Company. To accomplish both goals, ditches were needed, some to remove surface water and expose the fertile soil, others as waterways for flatboats loaded with logs to reach navigable streams elsewhere; some did both. In all, Washington made seven trips to the swamp, where he personally surveyed many of the locations for the network of ditches. The first of these was a five-mile-long trench connecting the western edge of the swamp with Lake Drummond; that trench, to no one's surprise, bears Washington's name. Subsequent ditches added about 125 miles to the network.

To further the shipment of logs, work started in 1793 on a canal that would connect Chesapeake Bay and Albemarle Sound, a project Colonel William Byrd II had proposed some sixty-five years earlier. In 1805, after twelve long years of toil, the twenty-two-mile-long canal was finished; in the next century, it would become a link in the Intracoastal Waterway. The canal also ended the flow of water eastward, which turned a large percentage of the original swamp—once a tract of more than a million acres—into tillable land.

Meanwhile, the massive volume of water was proving too much for the ditches to remove, and the Revolutionary War diverted the attention of Washington and his business partners for seven years. In 1785, when the ditching effort stalled, the partners hoped to revitalize the project with Dutch workers, but the prospect never materialized. For the most part, slaves constructed the ditches and canal under what were at best miserable working conditions. The drainage plan ultimately failed, but the ditches were effective enough to alter the swamp's hydrology to the point where they produced unwanted changes in the plant communities, not the least of which was an expansion of red maple. In contrast, the logging operations were too successful, at least in ecological terms, and likewise transformed the forest vegetation. Because of the straight-grained and durable qualities of its wood, bald cypress was the primary target, but tupelo and Atlantic white cedar also fell to the axe. The swamp forests have yet to fully recover from these disturbances.

Washington sold his shares in the Dismal Swamp Company in 1795, but they reverted to his estate—he died in 1799—when the buyer forfeited for lack of payment. Logging continued under various corporate entities through the next century. In 1909, Union Camp, a pulp and paper company, acquired the property and continued logging operations; the last stands of virgin timber were cut in the 1950s. Likely because of diminishing returns, Union Camp donated its holdings via transfer through The Nature Conservancy to the government in 1973; the property then became the core of the 112,000-acre Great Dismal Swamp National Wildlife Refuge.

A final note: in 2013, a ditch completed in North Carolina's Dismal Swamp State Park was dedicated to Martha Washington. Far from draining the swamp, this one channeled water *into* an area where the older ditches had long ago altered the flow of water. The project restored the natural hydrology to about three thousand acres in the Great Swamp National Wildlife Refuge.

are up to 25 percent larger than their "parent" species. Likewise, the Dismal Swamp bog lemming, once considered as an endemic, has been discovered elsewhere in North Carolina. These aside, the mammal fauna otherwise represents one typical of swamps elsewhere in the region.

Dismal Swamp provides important habitat for migrating songbirds, some of which remain to nest, whereas others continue their northward journey. Some 179 species of birds visit the area, which includes 85 nesting species. Swainson's warbler and Wayne's black-throated green warbler are noteworthy among the seventeen species of nesting warblers. Preferred habitat for the latter, a subspecies with a disjunct distribution on the Coastal Plain, includes Atlantic white cedar, which has been over-cut, to the detriment of the birds. Red-cockaded woodpeckers, once nesting in pocosins with a pond pine overstory, now seem extirpated. While it is probable that ivory-billed woodpeckers once occurred in the swamp, ornithological surveys in the late 1800s did not record the species.

Dismal Swamp is also an archaeological site rich with a history of human occupation that began at the end of the Pleistocene. During the Archaic Period, which extended from about 9,000 to 3,500 years ago, marshes dominated the area that is now swampland. The flocks of waterfowl in the marshes offered a welcome source of food, and so began a tradition of hunting with weapons known as bolas.

Bolas consist of two or more stones, each weighing about six ounces, that were attached to cords of equal length; these were tied together with a central knot. (When spread out, the arrangement resembled spokes on a wheel without a rim.) They were deployed by holding the central knot and whirling the stone weights overhead, then hurling it into the path of a flying duck. In the hands of a skilled hunter, the "spokes" wrapped around and disabled the hapless bird. Hundreds of bola weights have turned up in the swamp, including at a site near Washington's Ditch. The stones were fashioned in several shapes, three of which have not been found elsewhere. Some stones have grooves for binding the cords, whereas those without grooves probably filled leather pouches attached to the cords. Hunters fashioned the stones by chipping away unwanted material, using harder rocks as hammers, then smoothing the surface with abrasive materials. All of the stones discovered in the swamp were

How Red Maples Invade

Many locations within the broad realm of the eastern deciduous forest biome have taken on a new look: red maple is spreading and becoming a major player in stands of oak and hickory where it once occurred less frequently. Its steady advance also includes sites within the Great Dismal Swamp.

Several adaptations endow red maples with the ability to expand, not the least of which is their favorable response to human influences, particularly fire suppression and disturbances. Red maple, with its thin bark, has little resistance to fire, which once limited its range to wetter locations where it faced competition from species such as cypress and white cedar. However, when logging removed the latter, red maple readily filled the void. Meanwhile, modern firefighting capabilities and a network of roadways that act as firebreaks curtailed the incidence of fires, which removed another force that once held red maple in check. Given these circumstances, red maples move almost weed-like into disturbed areas, not only those associated with logging but also those where roadsides and ditching operations have similarly scarred the landscape.

Red maples also produce and spread seeds more quickly than other trees, an advantage that enhances their ability to invade disturbed areas. Each spring, even before sprouting leaves, red maples bear clusters of crimson seeds weeks or even months before other trees begin their annual reproductive cycle.

The flattened maple seeds—"samaras" to botanists—come equipped with a single, wing-like sail that "whirlybirds" when falling to the ground, often causing the seed to land at a considerable distance from the parent tree. Red maples thus benefit from a reproductive head start in the race for a place to grow. Moreover, trees as young as four to ten years old produce seeds and soon yield crops of twelve thousand to ninety thousand seeds per year, of which 85 percent or more germinate. As a further advantage, the shade-tolerant seedlings can grow under the canopy of other trees.

As a species, red maple enjoys what ecologists characterize as a broad ecological amplitude, meaning it thrives in a wide range of conditions. These include temperatures varying from the summer heat of Florida to the bitter cold of southern Canada, soils that vary from fertile to nutrient poor, and a moisture regime ranging from wetland to upland. Only an intolerance to fire and to stressful aridity thwarts its well-being. As such, red maple comes close to being an almost perfect species—a supergeneralist that occurs in any stage of forest succession, from a pioneer on disturbed sites to a century-old member of a climax community.

made from local rocks, except for two of iron ore (magnetite), whose nearest known source of use by other aboriginal people is in far-off Missouri!

Then, about thirty-five hundred years ago, the marshes gave way to accumulations of peat and the swamp forests we know today, and the waterfowl moved to better habitat elsewhere. Bolas likewise left the hunting scene, leaving stone-tipped arrows as the primary means for harvesting game. Whether the bola tradition evolved independently on site or was introduced by another culture remains unclear, although the latter seems more likely, given that bolas were used elsewhere in North America during the Archaic Period.

State and federal agencies protect large areas of the Dismal Swamp; these include state park and natural areas in North Carolina, as well as a 112,000-acre National Wildlife Refuge that straddles the border with Virginia. A relatively undisturbed area within the refuge was designated as a National Natural Landmark in 1973. Historically, the swamp served as a site where runaway slaves found safety and lived in small communities, whereas for others it was a link in the underground railroad north to freedom.

Pocosins: Shrub Bogs of the Southeastern Coastal Plain

Pocosins are not places for casual strolls, at least not without a sharp machete and briar-proof clothing. At ground level the vegetation is as thick and likely more so than any in North Carolina (Fig. 5-2, *top*). Common species in this all but impenetrable tangle include fetterbush, titi, coast pepperbush, wax myrtle, and gallberry holly. Other shrubs include honey cup, which features clusters of white, bell-shaped flowers that emit an appealing anise-like scent, along with blueberries and huckleberries, all members of the heath family, noted for its association with boggy sites with acidic soils. Most of the shrubs in the understory bear evergreen foliage. Curiously, the foliage of titi tends to be deciduous in northern pocosins but remains evergreen in southern locations.

Because of its sharp, half-inch prickles, laurel greenbriar quickly gains the attention of anyone braving a pocosin. The vines arise from large,

FIGURE 5-2. Tangles of impenetrable vegetation reach their zenith in pocosins (*top*). The edges of pocosins commonly arise abruptly next to pine savannas or other communities with more hospitable (at least for humans) prospects for access (*bottom*).

woody rhizomes and may grow as much as two inches per day, eventually capable of reaching lengths of five yards or more. The plants climb over the understory one shrub to another, gripping branches with their strong tendrils. It's not much of an exaggeration to regard the tough vines as the twine that wraps the pocosin package. The plants, also bearing evergreen foliage, produce shiny black berries, which take almost two years to ripen. Otherwise, few herbaceous plants find niches in pocosins.

If the understory in pocosins is thick, the overstory is not. Characteristically, the meager overstory consists of scattered pond pines, many of which feature crooked trunks and irregular tops. Its Latin name (*serotina*, meaning "late") refers to the delayed opening of its cones, a condition known as serotiny. This adaptation keeps the cones from dispersing their seeds until fire provides an environmental trigger; the tightly closed cones may remain on the tree for up to ten years, then open when fire sweeps through the pocosin. In contrast, the cones of most other pines open spontaneously when they mature, typically an annual event. Moreover, temperatures of more than three hundred degrees do not affect the viability of the seeds, which germinate in openings created by the fire. Pond pines also sprout needles directly from their trunks and limbs after fires. Some dendrologists consider pond pine as a subspecies of pitch pine, a species more commonly associated with sandy soils elsewhere (e.g., the Pine Barrens of New Jersey).

Peat commonly forms a deep layer over the underlying mineral soils in pocosins. The sponge-like nature of peat produces long hydroperiods, and while shallow pools of standing water frequently occur in pocosins, they seldom flood. Because of their saturation, the soils lack oxygen, which limits the abundance of bacteria and fungi necessary for the rapid decomposition of organic materials. The result is the nutrient-impoverished soil of peat, littered with leaves, twigs, and branches. Given the long hydroperiods in pocosins, peat fires occur infrequently (e.g., during droughts), but once ignited, they can burn for months. In doing so, however, they release nutrients that help establish a new generation of vegetation.

Pocosins lack a characteristic shape or size and may grade into other communities, but others rise with an almost wall-like facade (Fig. 5-2, *bottom*). Given the flat terrain of the Coastal Plain, a change of only a few inches in elevation may be enough to separate pocosins from adjacent vegetation. Pocosins develop in various locations but most commonly occur in Carolina bays (see below). Other sites include those with saturated soils, such as seeps. The word "pocosin" originated from an Algonquin term for "swamp on a hillside," which likely reflects their location on somewhat higher ground between streams where rainfall can accumulate on site.

Thanks to the dense vegetation of pocosins, relatively few studies have considered their biota, and, to date, no endemic species have come to light. Mammal populations include species found elsewhere, such as marsh rabbits, short-tailed shrews, and bobcats. Black bears rely heavily on pocosins for food throughout the Coastal Plain. In Bladen County, for example, pocosins in Carolina bays provide bears with more natural foods than they acquire in other types of habitat. Of these, the foliage and berries of laurel greenbriar top the list, and the shoots provide the bears with abundant protein. Bears also den in pocosins, where they bed in dense thickets of fetterbush and greenbriar, likely because of the lack of trees with cavities. Most pocosins in Carolina bays are too small to offer bears adequate escape cover during the hunting season; instead, they head for large swamps. Still, because of development, the continued presence of pocosins may be the key means of assuring that black bears remain abundant on the Coastal Plain. All told, however, pocosins support low densities and few species of mammals.

Pocosins provide food and cover for songbirds, especially those known as neotropical migrants—warblers, thrushes, and orioles, among others. Many of these birds forage on the insects that abound in pocosins. Once rested and sated with food, most species continue their journey northward, although some stay to nest in the dense vegetation, including some species with declining populations. Pocosins might offer suitable wintering habitat for American woodcock, but the saturated soils likely preclude the presence of enough earthworms to fulfill their dietary needs.

Among the amphibians and reptiles living in pocosins, Pine Barrens treefrogs and spotted turtles deserve brief mention, although both occur elsewhere on the Atlantic Coastal Plain. Because of their discontinuous distribution, Pine Barrens treefrogs come the closest to representing an endemic species; isolated populations occur in New Jersey and the Florida Panhandle, where the range spills over into a small area in Alabama. Otherwise, these colorful treefrogs remain closely associated with pocosins in North and South Carolina. The secretive species can best be found during the breeding season, April to July, by zeroing in on males as they voice a rapid succession of a nasal "quonk-quonk-quonk"; males call from the ground, the foliage of shrubs, or from water, which adds to the difficulty of finding one.

Spotted turtles, only three to five inches in length, bear distinctive yellow or orange spots on their heads and scutes. At hatching, most have just one spot per scute, but older individuals may have one hundred or more of these markings on their carapace (upper shell), prompting the name "polka-dot turtle." The spots may provide camouflage for the turtles, given their habit of seeking waters covered with duckweed, although some individuals inexplicably lack spotted carapaces. They occur in various types of habitat, including shallow wetlands in marshy pocosins. Males can be distinguished by their brown eyes and tan chin, whereas females have yellow or orange eyes and a pinkish or yellowish chin. Because of their small size and attractive markings, spotted turtles in some areas may be overexploited by the pet trade.

Carolina Bays: Mysterious Wetlands

Landforms known as Carolina bays have long baffled geographers and geologists, who have proposed various and sometimes strange explanations for the origin of these elliptical depressions (Fig. 5-3). Indeed, these landforms remained undefined until the advent of aerial photography in the 1930s. As interior wetlands, they of course bear no geographical similarity with shoreline embayments and instead were named for the bay trees that commonly grow at many of these sites.

Two physical features distinguish Carolina bays: they are aligned on a northeast–southwest axis, and a sandy rim typically rises along their southwestern edge. Counts vary, with some as high as five hundred thousand, but more likely ten thousand to twenty thousand Carolina bays dot

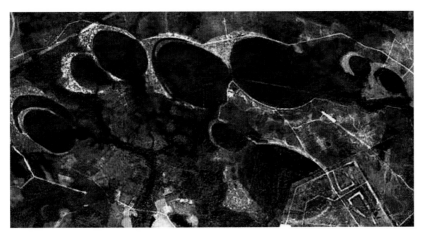

FIGURE 5-3. Carolina bays dot the landscape across large areas of the Atlantic Coastal Plain but nowhere occur more frequently than in Bladen County, North Carolina. Note that some depressions partially overlap others, which complicates explanations regarding their origin. Photo courtesy of North Carolina Division of Parks and Recreation.

the Coastal Plain from southern New Jersey to northern Florida. But nowhere are they as concentrated as in southeastern North Carolina, particularly in Bladen County, where about nine hundred occur.

Most Carolina bays have no connection with river systems or other types of surface water; instead, the interaction of precipitation and evapotranspiration alone governs their hydrology. Tannins add a tea-like color to the standing water in the wetter bays, although at White Lake State Park, the water originates from springs and therefore remains clear. Some Carolina bays cover several thousand acres, whereas others barely exceed an acre in size. They also vary in their vegetative development; cypress swamps fill some, as do marshes, but most in North Carolina are choked with pocosins, as described above. With the notable exception of Lake Waccamaw (see below), virtually all Carolina bays represent closed watersheds that lack well-defined drainage systems.

Among the more imaginative theories for their origins was one that proposed that fishes nesting in ancient seabeds excavated the depressions, but by far the most popular idea suggested that an immense

Hard Times for Red Bay

In the last century, Dutch elm disease and chestnut blight tragically imperiled the existence of their respective hosts, in the latter case with such severity that it has been likened to the extinction of the passenger pigeon. Now laurel wilt poses a similar threat to red bay and some of its relatives (e.g., sassafras and spicebush, but also to the commercial production of avocadoes). Indeed, estimates suggest that the disease has already killed some three hundred million trees, or about one-third of the red bay population, since it was discovered.

The pathogen, a fungus, is carried from tree to tree in the mouth parts of the redbay ambrosia beetle, a shiny-black, cylindrically shaped, wood-boring insect just one-sixteenth of an inch long. It takes only one female to infect a tree with the lethal fungus. Thereafter, the aggressive pathogen spreads rapidly through the host's vascular system, where the fungus blocks water transport, which causes wilting and the death of the tree usually within a few weeks (although the wilted leaves, by then brown, remain attached for months longer). The tiny holes bored by the beetles may be difficult to see without removing the bark, but a small, string-like cast—a mix of beetle excreta and fine sawdust—often provides a telltale clue to the burrow's location.

First noticed in 2002 in Georgia, the beetles likely arrived hidden in wood products shipped from the species' native home in Asia. Since then, the beetles and the destruction they cause have spread throughout the southeastern states from Florida to Texas, reaching North Carolina in 2011. As often is the case, the fungus does not harm trees in Asia, where the plants have coevolved defenses against its injurious effects.

The beetle's life cycle begins when a female finds a tree where she can bore a branched tunnel and deposit her eggs. She may not have mated and thus lays unfertilized eggs. These produce only males, which later become her mates. Her next batch of eggs, which are fertilized, then includes females, which disperse and begin another cycle; males in fact cannot fly. Redbay ambrosia beetles do not feed on wood but instead subsist on the fungus that they "farm" within the tree's vascular tissues—a symbiotic relationship of mutual benefit to both the beetles and the fungus they disperse to uninfected trees.

Red bay lacks commercial value as lumber, but it does offer cover for songbirds and other wildlife. More specifically, the larvae (caterpillar) of the Palamedes swallowtail butterfly rely almost exclusively on red bay as a primary food, and those of the spicebush swallowtail similarly depend on spicebush. Hence, both of these butterflies also may face an uncertain future because of laurel wilt. At present, no controls exist to halt the advance of the disease. However, because humans inadvertently spread the disease when they cut and move infected trees for firewood, the dead or dying trees should instead be burned on site.

meteor shower hit the eastern seaboard. As the story goes, the meteors hit at an angle; hence, their impact produced elliptical depressions with pushed-up rims at one end. However, no iron or nickel-laden fragments have ever turned up at these sites, which led to an alternate theory that a disintegrating comet gouged out the depressions and left no trace when the ice melted. A newer version suggests that the bays were formed by the secondary impacts of glacial ice boulders ejected by the strike of an extraterrestrial object. For the most part, however, geologists today generally agree that the depressions formed during the Pleistocene when dissolution of limestone in the underlying strata weakened the profile and the surface above subsided. The depressions then were modified by the dominant wind patterns, which created both the elliptical shapes and sandy rims that characterize the depressions. Even so, that scenario fails to explain how some bays came to partially overlap another (i.e., akin to a second footprint overlying another).

Carolina bays offer islands of biodiversity to the Coastal Plain. In particular, they provide amphibians of several species with essential breeding habitat and serve as informal refuges for an array of other wildlife, including black bears. Carolina bays also serve as breeding, foraging, and stopover (for nonbreeders) habitat for neotropical migrants, among them species currently declining because of habitat loss elsewhere in their range. The ground cover includes rare plants such as awned meadow beauty, distinguished by bright yellow anthers that form an abrupt angle where they attach to the stamens; these emit pollen through a tiny pore when buzzed by bumblebees.

Those Carolina bays that form temporary ponds offer breeding habitat for large numbers of frogs and toads of several species that otherwise live in the adjacent uplands. In such cases, the bays thus provide islands of wetland habitat suitable as breeding habitat, as well as a fish-free environment where tadpoles avoid a major source of predation. In season, the chorus of courting frogs and toads turns a bay into a vibrant concert hall. Pine Barrens treefrogs, ornate chorus frogs, and Brimley's chorus frogs are among the regular breeders in these settings. Another is the Carolina gopher frog, but unlike its relatives in the Deep South, this subspecies does not live in burrows excavated by gopher tortoises. Why not?

Simply because gopher tortoises do not occur in North Carolina. Instead, Carolina gopher frogs seek shelter in stump holes, rodent burrows, and root tunnels; they hunt at night from perches near the entrances to their hideaways.

Unfortunately, a Supreme Court ruling in 2002 no longer assures the protection of Carolina bays and other wetlands that lack natural connections to navigable bodies of water. As a result, many Carolina bays have been drained and cleared, as witnessed by large ovals of dark soil that stand out in aerial photos of croplands, each one a scar marking a lost resource.

Lake Mattamuskeet

Native Americans believed Lake Mattamuskeet formed after a peat fire that lasted for "thirteen moons" burned out a huge saucer-shaped depression, whereas some geologists later claimed it resulted from the impact of a meteor shower and thus presents another example of a Carolina bay. Whatever its origin might be, with an area once covering some 120,000 acres, Lake Mattamuskeet was and remains the largest natural lake in North Carolina. Translated as "dry dust," the Algonquin name for the site may reflect conditions at a time when drought parched the shallow basin and allowed winds to kick up the exposed soil. Droughts aside, the lake fills solely from surface runoff without any inflow from springs or streams; today the water depth in places may reach five feet, but, overall, it averages less than two feet.

Over time, the influx of sediments produced a lake bed of rich soil perhaps equaled only by the fertility of the Nile delta. The lake thus attracted agricultural interests as early as 1773 in a plan to drain the water into Pamlico Sound, followed by a similar plan in 1789, but neither of these produced results. In 1837, however, a canal was completed that diminished the lake to its current area of 40,100 acres and dropped the original depth of six to nine feet by more than half. Because the lake bed lies three to five feet below sea level, the canal required one-way gates to keep out salt water. As hoped, crops flourished on the newly drained lands.

Nonetheless, in the early twentieth century the lure of creating even more rich farmland produced several efforts to drain the remaining water. The outcome was a three-story pump house with the capacity of removing 1.25 million gallons of water per minute from the lake—completed in 1915, it was at the time the largest such facility in the world. However, the Great Depression ended the project, and in 1934 the lake property was sold to the federal government to establish the Mattamuskeet National Wildlife Refuge (Fig. 5-4). The Civilian Conservation Corps remodeled the pump house into a headquarters building and hunting lodge, topped by a 120-foot-high observation tower made from the smokestack. The lodge operated until 1974; the building was enrolled in the National Register of Historic Places in 1980. In 2006, the facility was transferred to the state of North Carolina.

The 50,180-acre Mattamuskeet National Wildlife Refuge provides habitat each winter for thousands of waterfowl, including northern pintails, green-winged teal, and sixteen other species of ducks, but especially for lesser snow geese and tundra swans. Wood ducks remain in residence year-round, many of which produce their broods in nesting boxes; each year a crew of refuge staff, interns, and volunteers check the boxes for usage and repairs. In the past, about one hundred thousand Canada geese overwintered at the refuge, but most of these birds later moved elsewhere—an event that triggered closing of the hunting lodge—whereas the winter population today numbers about five thousand. Ospreys and bald eagles are among the raptors at the refuge, which also offers excellent habitat for warblers and other neotropical migrants, as well as for numerous species of shorebirds during their seasonal passage up and down the Atlantic Flyway.

Several impoundments on the refuge enable drawdowns in which desirable food resources—seed-bearing vegetation, as well as invertebrates—flourish on soil exposed during the growing season; the units are reflooded in the fall just before the birds arrive. This season-specific technique, known as moist-soil management, benefits both waterfowl and shorebirds. In addition to its diverse avifauna, the refuge also offers attractive environments for black bears, otters, bobcats, deer, and other mammals.

FIGURE 5-4. Lake Mattamuskeet, in part a national wildlife refuge, provides seasonal habitat for large numbers of waterfowl, including tundra swans.

Tundra Swans in Winter

Tundra swans, previously called whistling swans, highlight the roster of 240 species of birds recorded at Mattamuskeet National Wildlife Refuge. (Some also overwinter at Pocosin Lakes and Pea Island National Wildlife Refuges.) The continental population of tundra swans overwinters in two widely separated regions, one in the Pacific Flyway, the other in the Atlantic Flyway, where sixty-five thousand to seventy-five thousand regularly occur in northeastern North Carolina. Of the latter, about twenty-five thousand birds occur on or near these refuges, but that was not always the case. Previously, Chesapeake Bay lay at the core of the species' winter range, but by the 1970s, after years of pollution finally devastated the bay's aquatic vegetation, most of these birds had moved south to northeastern North Carolina, where about 70 percent of the eastern population now winters. During this period, the birds also adopted a new form of feeding when they began foraging on winter wheat and other cereal crops in nearby fields.

Tundra swans overwintering in North Carolina nest in the far-off Canadian Arctic. Based on information provided by birds outfitted with transmitters, they fly a diagonal route to the Great Lakes, then continue to North Dakota and northward across Manitoba, Saskatchewan, and Alberta. Some veer off to nest on the western coast of Hudson Bay, whereas others head as far as the Mackenzie River delta in the Northwest Territories. The lengthy flight, which occurs in both spring and fall, includes layovers at several staging areas, where the birds rest and refuel before continuing their three-thousand-mile journey. Indeed, taken together, the spring and fall migrations occupy about half of their annual cycle, compared to just 20 percent of their time at wintering areas. These circumstances underscore the importance of conserving habitats where the birds stop at staging areas as much as those where they overwinter and breed.

Currently, only two states in the Atlantic Flyway—North Carolina and Virginia—hold hunting seasons for tundra swans, North Carolina having done so for more than three decades. North Carolina previously issued 5,000 permits each year but subsequently increased the number to 6,250 in 2017–18; hunters may harvest one swan per year and must complete a survey form at the end of the season. About half of the permit holders successfully fill the bag limit. The permit system takes into consideration annual assessments of breeding success and population estimates based on aerial surveys. Whereas these data fluctuate annually, the eastern population of tundra swans has greatly increased in recent decades, thereby indicating that hunting poses no threat to its welfare.

Still, Lake Mattamuskeet is not without problems, particularly in regard to water quality and its effects on aquatic food chains. Because of excessive nitrogen and phosphorous runoff from fertilizers, a process known as eutrophication accelerates the growth of algae to the point where water becomes so opaque that it inhibits light penetration. As a result, aquatic vegetation—wild celery, redhead grass, and sago pondweed, among others—no longer can maintain its photosynthetic activities and dies. Today most of the lake bottom is barren, thereby restricting the food supply for waterfowl dependent on leaves, tubers, and seeds for food. Eutrophication thus poses a significant environmental limitation to the lake's ability to maintain large numbers of waterfowl. As it happens, this situation is not the first time that water clarity posed an issue at the refuge. This occurred when the rooting activities of carp had so thoroughly muddied the water that the submersed vegetation died out, soon followed by the virtual elimination of bass and other game fish in the lake. However, massive efforts to remove the carp paid off, and the aquatic vegetation quickly recovered, after which the lake was restocked with game fish. Hopefully, similar results will again prevail when and if eutrophication can be curtailed. To that end, a management plan involving local stakeholders, as well as state and federal agencies, now tops the agenda at Mattamuskett National Wildlife Refuge.

Lake Waccamaw: Hot Spot for Speciation

Depending on one's point of view regarding the origin of Lake Mattamuskeet (see above), Lake Waccamaw ranks as either the largest or second largest Carolina bay in North Carolina. Measuring about 5.2 by 3.5 miles, the oval-shaped lake covers almost nine thousand acres at an average depth of 7.5 feet. Size aside, Lake Waccamaw also stands alone in comparison with virtually all other Carolina bays. Physically, sands cover about 70 percent of the lake bottom in a donut-like ring that surrounds a center of peat. Four streams feed the lake, each of which drains a swamp; hence, the inflow represents the tea-colored and acidic waters typically associated with the decomposition of tannin-rich organic matter. Lake Waccamaw serves as the source for the 140-mile-long Waccamaw River.

Together with the four streams, the river incorporates the lake into a drainage system quite unlike the watersheds of other Carolina bays that lack either inlets or outlets. Submerged vegetation includes waterweed and slender pondweed; the aquatic vegetation also features Cape Fear spatterdock. A dam spans the entrance to the Waccamaw River where it begins flowing downstream from the lake.

Geochemically, the acidic water flowing into Lake Waccamaw soon becomes more alkaline, thanks to the buffering action of low limestone cliffs on the north shore of the lake, and levels off at or near neutral (Fig. 5-5). These south-facing cliffs provide moist habitat for southern maidenhair, also known as Venus hair fern, which is the only naturally occurring population of the species in North Carolina and appears on the state's list of endangered species. In other places, a layer of the same limestone covers the lake bottom just under the sand and likewise buffers the acidic water. This calcareous deposit of ancient shells provides further evidence of the maritime history of the Coastal Plain.[1] More to the point, the lack of acidity facilitates greater diversity in the aquatic fauna at Lake Waccamaw, where, for example, the calcium-rich waters provide suitable habitat for eleven species of snails and eight species of mussels. By comparison, the acidic waters in blackwater lakes elsewhere limit the occurrence of various forms of life, especially snails and other shelled creatures (see "Blackwater Streams: A Coastal Specialty").

We should pause here to note that evolutionary biologists generally associate speciation—the development of new species from a common ancestor—with a scenario along the following lines: During a long period of time, the parental stocks remain reproductively isolated from other organisms with which they might otherwise breed, usually because they become physically separated. This typically occurs when some sort of physical barrier separates a larger population

---

1 In 2008, a swimmer discovered—with her toe—what proved to be the skull and jaws of a fossil whale embedded in an outcropping of limestone on the lake floor. The whale, in life about twenty feet long, dates to the late Pliocene or early Pleistocene (i.e., about 2.75 million years ago). Assigned to the extinct genus *Balaenula* but otherwise undescribed at the species level, the whale is a fossil relative of modern-day right whales and is on display at the visitors' center at Lake Waccamaw State Park.

FIGURE 5-5. An outcrop of limestone along the northern shore of Lake Waccamaw buffers the natural acidity of the lake's water—a geological feature that facilitated the development of a local fauna unlike that at other lakes on the Coastal Plain. The dark stain on the limestone reflects a period of high water. Note southern maidenhair ferns on the face of the outcrop.

into two smaller populations. Such barriers may be long gone (e.g., glacial retreat) and often difficult to describe in more than broad terms. Eventually, genetic changes alter the features in the isolated populations, including those that preclude successful breeding with any other population. In short, the heirs of this process can produce fertile offspring only by mating with others within their own population, which now represents a new species that may not occur anywhere else. At Lake Waccamaw, however, the combination of isolation and unique water chemistry, and not a physical barrier, allowed speciation to occur—and did so in a remarkably short time, given that the lake may be only fifteen thousand years old.

Lake Waccamaw, in fact, stands as the showcase for endemic species in North Carolina. These include three species of fishes, Waccamaw killifish, Waccamaw darter, and Waccamaw silverside, each in a different family. The Waccamaw silverside, known locally as a glass minnow (or skipjack, because of its habit of skipping across the surface), sometimes extends its range into the upper Waccamaw River during periods of high water; however, these incursions do not establish permanent populations, and the species otherwise occurs only in the lake. Distinctively marked with a silvery stripe on each side, these slim, almost transparent fish have large eyes and a jaw that juts sharply upward; they feed primarily on microcrustaceans and grow to about 2.5 inches in length. They breed at one year of age, with most dying shortly thereafter; nonetheless, they number in the millions and serve as the principal forage fish in the lake's aquatic food chain. Waccamaw silversides avoid aquatic vegetation and instead form schools that remain near the surface in open water over dark-bottomed sites. Because of its highly restricted distribution, the species appears on the federal list of threatened species.

A comparison between the Waccamaw darter and its close relative, the more widely occurring tessellated darter, from which it was likely derived, reveals strong genetic similarities between the two species. The two darters come into contact in the upper reaches of the Waccamaw River, but they nonetheless remain reproductively isolated, in part because of significant differences in their respective life histories. In particular, the elaborate courtship behavior of the Waccamaw darter has evolved to the point where it does not attract potential mates from the available pool of tessellated darters. North Carolina lists the Waccamaw darter as a threatened species (Fig. 5-6).

The Waccamaw killifish, the third of the trio of endemic fishes, likely evolved as a derivative of the banded killifish.[2] Between early spring and late summer, Waccamaw killifish remain in the lake, where they favor shallow water and the presence of aquatic vegetation. In winter, most of the killifish move to feeder streams or the headwaters of the Waccamaw River. North Carolina currently lists the Waccamaw killifish as a species of special concern.

When breeding, the cross-bar markings on the males assume an iridescent blue-green hue, as do their pelvic and anal fins. Thus dressed, breeding males temporarily defend circular territories about three feet in diameter from their rivals in adjacent territories. For a mental image, picture several hula hoops laid out on the lake floor, with a male killifish patrolling the rim of each hoop. When two males happen to meet, each elaborately displays his fins and gills in a face-to-face encounter, or they assume a side-by-side position and "beat" each other with their tails. These encounters soon end, and the rivals continue patrolling the perimeter of their respective territories, which they abandon two to three hours later.

Meanwhile, one of several females waiting nearby enters the territory of a particularly attractive (to her) male and, after a bit of stimulation, spawns an egg, which is immediately fertilized and adheres to sand grains or bits of plant material on the lake bottom. Mating completed, the male then resumes his patrol until another female enters his territory. This arrangement takes on certain characteristics of lek behavior, in which males display in an arena—the lek—surrounded by an audience of females, a form of courtship best known in birds such as prairie

2   Some confusion surrounds what appears to be a second population of Waccamaw killifish in Lake Phelps, located some 155 miles northeast of Lake Waccamaw, although this may be the result of escaped bait fish rather than a case of parallel evolution. Still, because of some minor physical differences, the taxonomic status of the "Phelps Lake" killifish has yet to be formally determined.

chickens. In the case of prairie chickens, however, only a few dominant males actually mate with the females that visit the arena, whereas most of the male killifish manage to gain mates.

Two other endemics, both snails, also highlight the uniqueness of the lake's fauna. These are the Waccamaw siltsnail and Waccamaw amnicola, but neither species has received much study. Each is listed by the state as a species of special concern. Two other mollusks, mussels in this case, were once considered as part of the lake's endemic fauna: the Waccamaw fatmucket and Waccamaw spike. However, this distinction has not held up to the scrutiny of molecular genetics, as both bear the same genetic signature of other species occurring in drainages outside Lake Waccamaw. Nonetheless, these and other native species of freshwater mussels remain at some level of risk, including those formally designated as endangered, largely because of their intolerance to diminished water quality.

The marvel of Lake Waccamaw faces a serious challenge from hydrilla, an invasive aquatic weed introduced from its native range in Africa and Asia. Once valued as an aquarium plant, it gained a foothold in North

America after some unwanted sprigs were thoughtlessly dumped in ditches in Florida; it now occurs widely across much of the United States. Hydrilla forms dense, impenetrable mats that clog the surface of infested lakes, making boating all but impossible and ruining habitat for game fishes and other forms of aquatic life. Its stems can fill the water column up to depths of twenty-five feet and completely crowd out native vegetation. Under ideal conditions, hydrilla may double its biomass every two weeks. It commonly spreads from lake to lake when stem fragments attached to boats break free and establish new populations, as likely happened at Lake Waccamaw. Hydrilla can be controlled, if not eliminated, with herbicides, and a ten-year program to do so at Lake Waccamaw has so far produced encouraging results.

The Saga of Red Wolves

Historically, the southeastern United States was home for red wolves, a slightly smaller and reddish version of the more familiar gray (or timber) wolf (Fig. 5-7). Their range originally extended across a broad swath from Pennsylvania to Texas and included all of North Carolina. But in a fate similar to that of their larger cousins to the north and west, red wolves disappeared as Europeans and their descendants settled in North America, bringing with them guns, traps, and, later, poisons along with little tolerance for predators. Western civilization and predators, especially large ones, just don't mix well (see the Afterword). Prime habitat for the wolves also disappeared, leaving only marginally suitable sites available for the few wolves that remained. By the mid-twentieth century, red wolves survived only in a few coastal counties in southeastern Texas and adjacent Louisiana. By then, a new threat had emerged: coyotes had arrived and readily bred with the wolves to the extent that the genetic integrity of the wolves was compromised by a process popularly known as genetic swamping and more technically as introgression. Simply put, ever-fewer red wolves remained free of coyote genes and traits, and with such a small population, they would soon disappear altogether as a distinct species. Extinction loomed.

Red wolves thus were listed as endangered in 1967 under the aegis of the forerunner to the Endangered Species Act of 1973.[3] As the first step to preserve the species, the remaining wild wolves were captured and sorted—those lacking any visible coyote characteristics became the core of a captive breeding program at a zoo in Tacoma, Washington. As hoped, the captive population grew to the point where selected individuals could be released at protected sites within the original range of the species. The pilot program started in 1978 when a pair of red wolves was released on an undeveloped barrier island at Cape Romaine National Wildlife Refuge in South Carolina. The transplants were fruitful; in 1987 four pairs of their descendants were transported and released at Alligator River National Wildlife Refuge in North Carolina. The program at Cape Romaine National Wildlife Refuge was phased out in 2005, when the remaining wolves were released at Alligator River National Wildlife Refuge or placed in captive breeding facilities. In 1992, red wolves were also released at Great Smoky Mountains National Park, but the program ended in 1998, and the survivors were added to the population in North Carolina. Currently, another two hundred or so red wolves remain in about forty captive breeding facilities.

By 2006, about 130 red wolves roamed the swampy wetlands in and around Alligator River National Wildlife Refuge (i.e., the Albemarle-Pamlico Peninsula), then declined to fewer than 40, in part because of roadkills and gunshot mortality. Recent estimates suggest that even fewer persist in the wild, making red wolves the rarest canid in the world. As might be expected, those wolves venturing onto private lands were not always welcome, not only by owners of pets and livestock but also by deer hunters. But more trouble loomed: the expanding range of coyotes now included North Carolina and, indeed, all of the states east of the Mississippi River. With the two canids again in contact, genetic

---

3   The 1973 act extends protection to subspecies, as well as to "any distinct population segment of any species . . . which interbreeds when mature" whose survival is in jeopardy. Stated otherwise, the act treats both of these groups as a "species" for the purposes of legal protection, restoration, and recovery.

FIGURE 5-7. Controversy has surrounded the status of red wolves as an endangered species. Today, the only remaining wild population hangs on at federal refuges and surrounding areas of privately owned land in northeastern North Carolina.
Photo credit: Becky Harrison, U.S. Fish and Wildlife Service.

introgression once more threatened the genetic integrity of red wolves. This, in turn, cast doubts about their status as an endangered species: Were the animals still "pure" enough to warrant legal protection from the Endangered Species Act? Some critics went even further and proposed that red wolves never were a valid species in the first place (e.g., they had originated as hybrids of gray wolves and coyotes). Indeed, the genus *Canis* is genetically "messy," given that all of its members can produce fertile offspring—"coydogs," for example—which belies the traditional definition of a species.

Meanwhile, the U.S. Fish and Wildlife Service devised a plan that might minimize contact between red wolves and coyotes. When enacted, the plan created a buffer zone occupied by a resident population of sterilized coyotes, which, because of their territorial behavior, became "place holders" that kept other coyotes out of an area occupied only by red wolves. Thus freed of contact with fertile coyotes, the wolves could maintain their own territories and then reproduce solely with others of their own kind. It worked: as of 2015, the genetic composition of red wolves in North Carolina reflected less than 4 percent coyote ancestry. To maintain that level of genetic integrity, however, required constant monitoring of the buffer zone to determine if the place holders were still present; if not, then any newcomers would have to be captured and sterilized as replacements. As a result, red wolves fall into a category known as conservation reliant, which means that the species must be managed to survive. Currently, however, so few wolves remain in the North Carolina population that the novel technique lost its effectiveness; hence, biologists no longer maintain the hybrid zone.

Yet another factor emerged in 2018 when red wolf genes turned up in a small canid population on Galveston Island, Texas, some forty years after the wolves had been declared extinct in the wild. This revelation indicated that red wolf genes still persisted after many generations of interbreeding with coyotes (i.e., genetic swamping had not erased the wolves' genetic identity). In theory, the discovery also might provide a means to bolster the gene pool of the wolves in North Carolina and perhaps others that might someday be released to establish wild populations elsewhere.

The matter of local resistance to red wolves remains an issue. In 2015, the North Carolina Wildlife Resources Commission requested that the federal management program withdraw from private lands. The U.S. Fish and Wildlife Service accordingly proposed to reduce the area where sterile coyotes were limiting introgression; however, a federal judge ordered a review, which blocked the move. Similarly, landowners were granted permits to shoot any wolves wandering onto private property, but a federal court subsequently rejected the policy.

One point of contention focused on the diet of red wolves: To what extent were they killing deer? Most of prey taken by the last wild population of red wolves—those in Texas and Louisiana—consisted of nutria, swamp and cottontail rabbits, and, to a lesser extent, rodents such as marsh rice rats and muskrats. Deer, however, were not included, but this changed after the wolves were released in North Carolina. There the wolves prey on white-tailed deer, as well as on raccoons and rabbits, although the composition of each varies among the packs in relation to the availability of the prey. Coyotes, too, prey on deer in the same area but not necessarily in the same proportion as the wolves. Elsewhere in the southeastern United States, coyotes prey heavily on deer, which suggests that similar losses might occur in North Carolina even if the red wolf population were to fade away. In fact, it might be argued that red wolves once checked the size and range of coyote populations, which, in turn, actually limited the toll of deer.

All of these matters merged when opponents of the reintroduction program convinced Congress to allocate almost $400,000 to address a single question: Is the red wolf a valid species and hence subject to the provisions of the Endangered Species Act? After a year of study, the National Academies of Sciences, Engineering, and Medicine issued their findings. The exhaustive review included behavioral traits, skull and other body measurements, and genetics, all leading to the conclusion that the red wolf of today represents a distinct species. Based on evidence in the report, the species diverged from coyotes at least fifty-five thousand years ago and lacks a close genetic relationship with gray wolves, as would be expected of a subspecies. Strong as it may be, the conclusion has yet to be translated into an updated and comprehensive plan to maintain a wild population of red wolves.

READINGS AND REFERENCES

"A Vast Body of Mire and Nastiness"

Bottoms, E., and F. Painter. 1979. Evidence of aboriginal utilization of the bola in the dismal swamp area. Pages 44–56 *in* The Great Dismal Swamp (P. K. Kirk Jr., ed.). University Press of Virginia, Charlottesville.

LeGrand, H. E., Jr. 2000. The natural features of Dismal Swamp State Natural Area, North Carolina. Pages 41–50 *in* The Natural History of the Great Dismal Swamp (R. K. Rose, ed.). Suffolk-Nansemond Chapter, Izaak Walton League of America, Suffolk, Va.

Levy, G. F. 1991. The vegetation of the Great Dismal Swamp: a review and overview. Virginia Journal of Science 42:411–417.

Levy, G. F. 2000. The lake of the Dismal Swamp. Pages 33–40 *in* The Natural History of the Great Dismal Swamp (R. K. Rose, ed.). Suffolk-Nansemond Chapter, Izaak Walton League of America, Suffolk, Va.

Meanley, B. 1979. An analysis of the birdlife of the dismal swamp. Pages 261–276 *in* The Great Dismal Swamp (P. W. Kirk Jr., ed.). University Press of Virginia, Charlottesville.

Oaks, R. Q., and D. R. Whitehead. 1979. Geologic setting and origin of the Dismal Swamp, southeastern Virginia and northeastern North Carolina. Pages 1–24 *in* The Great Dismal Swamp (P. W. Kirk Jr., ed.). University Press of Virginia, Charlottesville.

Sayers, D. O. 2014. A Desolate Place for a Defiant People: The Archaeology of maroons, Indigenous Americans, and Enslaved Laborers in the Great Dismal Swamp. University Press of Florida, Gainesville.

Wagner, W. H., Jr., and L. J. Musselman. 1979. Log ferns (*Dryopteris celsa*) and their relatives in the Dismal Swamp. Pages 127–139 *in* The Great Dismal Swamp (P. W. Kirk Jr., ed.). University Press of Virginia, Charlottesville.

Watts, B. D., B. J. Paxton, and F. M. Smith. 2011. Status and habitat use of Wayne's black-throated green warbler in the northern portion of the South Atlantic Coastal Plain. Southeastern Naturalist 10:333–344.

Webster, W. D., N. D. Moncrief, B. E. Gurshaw, et al. 2009. Morphometric and allozymic variation in the southeastern shrew (*Sorex longirostris*). Jeffersoniana 21:1–13. [Determined a wider distribution for *S. l. fisheri* than previously supposed.]

Webster, W. D., A. P. Smith, and K. W. Markham. 1992. A noteworthy distributional record for the Dismal Swamp bog lemming (*Synaptomys cooperi helaletes*) in North Carolina. Journal of the Elisha Mitchell Scientific Society 108:89–90.

Pocosins: Shrub Bogs of the Southeastern Coastal Plain

Clarks, M. K., D. S. Lee, and J. B. Funderburg. 1985. The mammal fauna of Carolina bays, pocosins, and associated communities in North Carolina: an overview. Brimleyana 11:1–38.

Landers, J. L., R. J. Hamilton, A. S. Johnson, and R. L. Marchinton. 1979. Foods and habitat of black bears in southeastern North Carolina. Journal of Wildlife Management 43:143–153.

Lee, D. S. 1986. Pocosin breeding bird fauna. American Birds 40:1263–1273.

Mamo, L. B., and E. G. Bolen. 1999. Effects of area, isolation, and landscape on the avifauna of Carolina bays. Journal of Field Ornithology 70:310–320. [Pocosins occupied the sites in this study.]

Richardson, C. J., ed. 1981. Pocosin Wetlands: Integrated Analysis of Coastal Plain Freshwater Bogs in North Carolina. Hutchinson Ross, Stroudsburg, Pa.

Richardson, C. J. 1983. Pocosins: vanishing wastelands or valuable wetlands? BioScience 33:626–633.

Sharitz, R. R., and J. W. Gibbons. 1982. The ecology of southeastern shrub bogs (pocosins) and Carolina bays: a community profile. FWS/OBS-82-04, Division of Biological Services, U.S. Fish and Wildlife Service, Washington, D.C.

Carolina Bays: Mysterious Wetlands

Brooks, M. J., B. E. Taylor, and A. H. Ivester. 2010. Carolina bays: time capsules of culture and climate change. Southeastern Archaeology 29:146–163.

Carver, R. E., and G. A. Brook. 1989. Late Pleistocene paleowind directions, Atlantic Coastal Plain, USA. Palaeogeography, Palaeoclimatology, Palaeoecology 74:205–216.

Howell, N., A. Krings, and R. R. Braham. 2016. Guide to the littoral zone vascular plants of Carolina bay lakes (USA). Biodiversity Data Journal 4:e7964. doi:10.3897/BDJ.4.e7964.

Johnson, D. W. 1942. The Origin of the Carolina Bays. Columbia University Press, New York.

Prouty, W. F. 1952. Carolina bays and their origin. Geological Society of America Bulletin 63:167–224.

Ross, T. E. 1987. A comprehensive bibliography of the Carolina bays. Journal of the Elisha Mitchell Scientific Society 103:28–42. [Includes early theories of how the bays originated.]

Savage, H., Jr. 1982. The Mysterious Carolina Bays. University of South Carolina Press, Columbia.

Sharitz, R. R. 2003. Carolina bay wetlands: unique habitats of the southeastern United States. Wetlands 23:550–562.

Zamora, A. 2017. A model for the geomorphology of the Carolina bays. Geomorphology 282:209–216. [Suggests that boulders of ice, launched by an extraterrestrial object hitting the Laurentide Ice Sheet, gouged out the bays (i.e., secondary impacts).]

Lake Mattamuskeet

Anonymous. 2008. Mattamuskeet comprehensive conservation plan. U.S. Fish and Wildlife Service, Atlanta, Ga.

Bellrose, F. C. 1955. Housing for wood ducks. Illinois Natural History Survey Circular 45.

Cahoon, W. G. 1953. Commercial carp removal at Lake Mattamuskeet, North Carolina. Journal of Wildlife Management 17:312–317.

Forrest, L. C. 1999. Lake Mattamuskeet, New Holland, Hyde County. Acadia, Charleston, S.C.

Fredrickson, L. H. 1996. Moist-soil management: 30 years of field experimentation. International Waterfowl Symposium 7:168–177.

Moorman, M. C., T. Augspurger, J. D. Stanton, and A. Smith. 2017. Where's the grass? Submerged aquatic vegetation and declining water quality in Lake Mattamuskeet. Journal of Fish and Wildlife Management 8:401–417.

Lake Waccamaw: Hot Spot for Speciation

Casterlin, M. E., W. W. Reynolds, D. G. Lindquist, and C. G. Yarbrough. 1984. Algal and physiochemical indicators of eutrophication in a lake harboring endemic species: Lake Waccamaw, North Carolina. Journal of the Elisha Mitchell Scientific Society 100:83–103.

Davis, J. R., and D. E. Louder. 1969. Life history and ecology of *Menidia extensa*. Transactions of the American Fisheries Society 98:466–472.

Frey, D. G. 1951. Fishes of North Carolina's bay lakes and their intraspecific variation. Journal of the Elisha Mitchell Scientific Society 67:1–44.

Hubbs, C. L., and E. C. Raney. 1946. Endemic fish fauna of Lake Waccamaw, North Carolina. Miscellaneous Publications, Museum of Zoology, University of Michigan, No. 65. Ann Arbor.

Jacono, C. C., M. M. Richerson, V. H. Morgan, and I. A. Pfingsten. 2018. *Hydrilla verticillata*. USGS Nonindigenous Aquatic Species Data Base, Gainesville, Fla.

Lindquist, D. G., J. R. Shute, and P. W. Shute. 1981. Spawning and nesting behavior of the Waccamaw darter, *Etheostoma perlongum*. Environmental Biology of Fishes 6:177–191.

McCartney, M. A., and F. S. Barreto. 2010. A mitochondrial DNA analysis of the species status of the endemic Waccamaw darter, *Etheostoma perlongum*. Copeia 2010:103–113.

McCartney, M. A., A. E. Bogan, K. M. Sommer, and A. E. Wilbur. 2016. Phylogenetic analysis of Lake Waccamaw freshwater mussel species. American Malacological Bulletin 32:109–120.

Shute, J. R., P. W. Shute, and D. G. Lindquist. 1981. Fishes of the Waccamaw River Drainage. Brimleyana 6:1–24.

Shute, P. W., D. G. Lindquist, and J. R. Shute. 1983. Breeding behavior and early life history of the Waccamaw killifish, *Fundulus waccamensis*. Environmental Biology of Fishes 8:293–300.

Shute, P. W., J. R. Shute, and D. G. Lindquist. 1982. Age, growth, and early life history of the Waccamaw darter, *Etheostoma perlognum*. Copeia 1982:561–567.

Stager, J. C., and L. B. Cahoon. 1987. Age and trophic history of Lake Waccamaw. Journal of the Elisha Mitchell Scientific Society 103:1–13.

The Saga of Red Wolves

Beeland, T. D. 2013. The Secret World of Red Wolves: The Fight to Save North America's Other Wolf. University of North Carolina Press, Chapel Hill.

Bohling, J. H., J. Dellinger, J. M. McVey, et al. 2016. Describing a developing hybrid zone between red wolves and coyotes in eastern North Carolina. Evolutionary Applications 22:74–86.

Bohling, J. H., and L. P. Waits. 2015. Factors influencing red wolf–coyote hybridization in eastern North Carolina. Biological Conservation 184:108–116.

Gese, E. M., F. F. Knowlton, J. R. Adams, et al. 2015. Managing hybridization of a recovering endangered species: the red wolf *Canis rufus* as a case study. Current Zoology 61:191–205. [Presents the results of the "place holder" management program.]

Heppenheimer, E., K. Brzeski, R. Wooten, et al. 2018. Rediscovery of red wolf ghost alleles in a canid population along the American Gulf Coast. Genes 9(12):618, http://doi/org/10.3390/genes9120618.

Hinton, J. W., A. K. Ashley, J. A. Dellinger, et al. 2017. Using diets of *Canis* breeding pairs to assess resource partitioning between sympatric red wolves and coyotes. Journal of Mammalogy 98:475–488.

Kilgo, J. C., H. S. Ray, C. Ruth, and K. V. Miller. 2010. Can coyotes affect deer populations in southeastern North America? Journal of Wildlife Management 74:929–933.

McVey, J. M., D. T. Cobb, R. A. Powell, et al. 2013. Diets of sympatric red wolves and coyotes in northeastern North Carolina. Journal of Mammalogy 94:1141–1148.

National Academies of Sciences, Engineering, and Medicine. 2019. Evaluating the Taxonomic Status of the Mexican Gray Wolf and the Red Wolf. National Academies Press, Washington, D.C.

Riley, G. A., and R. T. McBride. 1972. A survey of the red wolf (*Canis rufus*). Scientific Wildlife Report No. 162, U.S. Fish and Wildlife Service, Washington, D.C. [Describes wolf diets in Texas and Louisiana.]

Infobox 5-1: George Washington's Ditch

Brown, A. C. 1970. The Dismal Swamp Canal. Norfolk County Historical Society, Chesapeake, Va.

Royster, C. 1999. The Fabulous History of the Dismal Swamp Company: A Story of George Washington's Times. Alfred A. Knopf, New York.

Sayers, D. O. 2014. A Desolate Place and a Defiant People: The Archaeology of Maroons, Indigenous Americans, and Enslaved Laborers. University Press of Florida, Gainesville.

Infobox 5-2: How Red Maples Invade

Abrams, M. D. 1998. What explains the widespread expansion of red maple in eastern forests? BioScience 48:355–364.

Christensen, N. L. 1977. Changes in the structure, pattern, and diversity associated with forest maturation in Piedmont, North Carolina. American Midland Naturalist 97:176–188.

Infobox 5-3: Hard Times for Red Bay

Hall, D. W. 2018. Palamedes swallowtail, *Papilo palamedes* (Drury) (Insecta: Lepidoptera: Papilionidae). University of Florida Featured Creatures. Extension Publication No. EENY 60. http://entnemdept.ufl.edu/creatures/bfly/palamedes_swallowtail.htm.

Hanula, J. L., A. E. Mayfield III, S. W. Fraedrich, and R. J. Rabaglia. 2008. Biology and host associations of redbay ambrosia beetle (Coleoptera: Curculionidae: Scolytinae), exotic vector of laurel wilt killing redbay trees in the southeastern United States. Journal of Economic Entomology 101:1276–1286.

Hughes, M. A., J. J. Riggins, F. H. Koch, et al. 2017. No rest for the laurels: symbiotic invaders cause unprecedented damage of southern USA forests. Biological Invasions 19:2143–2157.

Kendra, P. E., W. S. Montgomery, J. Niogret, and N. D. Epsky. 2013. An uncertain future for American Lauraceae: a lethal threat from redbay ambrosia beetle and laurel wilt disease (a review). American Journal of Plant Sciences 4:727–738.

Infobox 5-4: Tundra Swans in Winter

Crawley, D. R., Jr., and E. G. Bolen. 2002. Effect of tundra swan grazing on winter wheat in North Carolina. Waterbirds 25 (special publication 1):162–167.

Howell, D. L. 2018. 2017–18 North Carolina tundra swan season and survey report. North Carolina Wildlife Resources Commission, Raleigh.

Limpert, R., and S. L. Earnst. 1994. Tundra swan (*Cygnus columbianus*). The Birds of North America, No. 89 (A. Poole and F. Gill, eds.). Academy of Natural Sciences, Philadelphia, Pa., and American Ornithologists' Union, Washington, D.C.

Petrie, S. A., and K. L. Wilcox. 2003. Migration chronology of eastern-population tundra swans. Canadian Journal of Zoology 81:861–870.

Serie, J. R., D. Luszcz, and R. V. Raftovich. 2002. Population trends, productivity, and harvest of eastern population tundra swans. Waterbirds 25 (special publication 1):32–36.

Fire in longleaf pine forests is like rain in a rain forest.

—LAWRENCE S. EARLEY, *Looking for Longleaf: The Fall and Rise of an American Forest*

VI

Coastal Uplands

A Vast Plain of Longleaf Pine

More than two centuries ago, William Bartram in his classic *Travels* marveled at a "vast plain . . . of the great longleaved pine" in which the ground was "covered with grass, interspersed with an infinite variety of herbaceous plants." With these few words, the famed traveler-botanist had described one of the most species-rich communities in North America. At the time, it was also one of the continent's largest communities, extending over a broad arc from southern Virginia to eastern Texas. George Washington on his "Southern Tour" of the newly founded United States noticed another facet of the community as he crossed North Carolina's Coastal Plain between New Bern and Wilmington in 1791. With a planter's eye for the land, the nation's first president recorded in his diary that a "course [*sic*] grass [had] sprung since the burning of the woods." Washington had observed the remarkable adaptation of what is today known as wiregrass, a fire-dependent associate in the understory of longleaf pine forests. In time, the association between fire and diversity in these communities would become a hallmark in the annals of natural history.[1]

---

1   Esteem for the pines is further reflected by membership in the Order of the Longleaf Pine. Issued by the governor, the recognition is the highest award for service to North Carolina.

Longleaf pine itself is fire dependent, as reflected in its growth and development. The cycle begins when fires clear the forest floor of needles and other debris, thereby uncovering the underlying mineral soil, on which the seeds germinate. These and subsequent fires also reduce the threat of brown-spot needle blight, a serious fungal disease harbored in the ground litter; fires kill the spores that infect and defoliate the early growth of longleaf and other pines. Referred to as "cool fires," the flames of these low-intensity ground fires rise about knee high, quite unlike the intensely hot conflagrations sweeping through forest canopies and killing nearly all vegetation in their path (e.g., the horrific fires that swept across Yellowstone National Park in 1988).

Quite unlike those of other pines, the seedlings of longleaf pine develop as a grass-like clump of needles that surround the plant's growing point (terminal bud) and protect it from all but the hottest fires (Fig. 6-1, *left*). During this "grass stage," which typically lasts for three to six years, the central stem of the seedlings does not grow vertically; instead, the plants develop taproots that extend to depths of nearly ten feet to ensure access to water in sandy soils. In all, most fires occurring at this stage of development pose little threat to the seedlings, whereas they do control the establishment and growth of would-be competitors whose terminal buds lack similar protection (e.g., hardwoods, especially oaks).

Then, with remarkable speed, the central stem, often likened to a "white candle," spurts upward, cloaked with needles and the beginnings of lateral branches. With favorable conditions, the seedlings grow twelve to thirty-five inches per year. The result is a six-to-ten-foot column-shaped tree that resembles a bottlebrush; for a time, fires can damage the young trees at this stage, but the rapid elongation of the stem quickly elevates the growing point beyond the reach of ground fires (Fig. 6-1, *right*).

When mature, the straight trunks of longleaf pines may exceed one hundred feet in height and, true to their name, produce needles of twelve to eighteen inches in length; these appear in bundles ("fascicles") of three. Because of their strong, straight trunks, the Royal Navy in Colonial times reserved some of the taller trees for ship masts; these were blazed with an arrow to indicate a "king's tree," which supposedly accounts for the

FIGURE 6-1. Initially, the seedlings of longleaf pine appear as tufts of grass (*left*). In a spurt of growth three to six years later, these develop into saplings that resemble bottlebrushes (*right*) before assuming the typical profile of a pine.

name of Kingstree, South Carolina. Their cones commonly reach lengths of eight to twelve inches, which, along with their needles, are the largest of any pine growing east of the Mississippi River. Longleaf pines begin producing cones when twenty to thirty years old, each housing fifty to sixty seeds. The slow-growing trees may live for three hundred years. As the pines mature, their bark thickens into armor, which protects the underlying cambium from all but the most severe fires. Air-filled cells in the corky bark—imagine fire-resistant bubble wrap—provide efficient insulation against temperatures that commonly exceed 1,000 degrees Fahrenheit, whereas the vital cambium cannot survive 140 degrees. Based on tests conducted with a propane torch, the bark of longleaf pine is in fact nearly twice as heat resistant as bark of the same thickness from several species of hardwood (e.g., red maple).

Under natural conditions, longleaf pines develop in even-aged stands. Typically, this occurs because a rare "hot fire" has indeed killed a stand of mature trees and thereby provided a litter-free, unshaded area where a new generation of seedlings can develop. Extended droughts

Naval Stores: Wooden Ships and Tar Heels

Extensive forests of longleaf pine once covered the Coastal Plain, where their products—primarily naval stores—formed the commercial backbone of the regional economy in the early years of North Carolina's history. Historian Alan D. Watson has calculated that 52,708 barrels of tar, 8,627 barrels of turpentine, and 919 barrels of pitch passed through Port Brunswick in a single year (1767–68); now gone, this once busy port lay south of present-day Wilmington. Only later, in the first half of the nineteenth century, would "King Cotton" become an economic force in the region. The distribution of both commodities, however, depended on the Cape Fear River.

True to their name, naval stores, which include turpentine, pitch, and tar obtained from longleaf pine, served the needs of wooden ships, largely by treating ropes and canvas against rot, as well as by sealing the seams and joints between planking. Harvesting began when the trees were "boxed" by a series of slashes—each "V" shaped—from which the resin oozed into collection pots located at the tip of the cut in a process somewhat similar to the way maple sap is collected. New cuts were added when the flow diminished, so that the site eventually resembled upside-down sergeants' stripes. The same sticky resin sought by the turpentiners guards the nest holes of red-cockaded woodpeckers (see the main text).

The resin, at this point called crude turpentine, was distilled to produce turpentine spirits and left behind a residue of rosin, whereas the trees themselves were burned in kilns to yield tar. The latter, when boiled, produced an even thicker substance known as pitch; three barrels of tar yield two of pitch. Naval stores also seeped into our everyday language and history. Sailors of the day stained their hands on the treated ropes and hence were dubbed "tars," and because North Carolina's militiamen held their ground in the thick of battle, their tenacity earned them the distinction as "tar heels," now applied more widely to the state's citizenry and to the athletic teams at Chapel Hill. Moreover, access to naval stores was among the reasons England fought to maintain its colonies in America. Britannia "ruled the waves" with its fleet; hence, the loss of the primary source of pitch and tar posed a serious threat to the continued existence of its far-flung empire.

The long inland reach of the Cape Fear River offered a cheap means of conveying the heavy barrels to port. Flatboats and rafts rode the current downstream, guided by crews handling long poles or oars mounted at each end of the craft. Progress was slow but far cheaper than passage overland on the few rutted tracks that passed for roadways (which necessarily were used when low water made the river unnavigable). The watermen on these crafts received three gallons of rum per week to supplement their salary—compensation that they perhaps valued as much as their wages.

The industry, although diminished by overcutting of the pine forests and the advent of steel-hulled, steam-powered ships, continued into the early twentieth century before it ended when petroleum-based mineral spirits claimed the turpentine market. Scant evidence of the once viable industry remains, although a few box-scarred trees still survive in some locations, among them the campus at UNC Wilmington. A kiln at Moore's Creek National Battleground and others at Goose Creek State Park also remind visitors of a bygone time when naval stores were an economic force.

Today, longleaf forests provide another crop—pine needles, raked and baled, are sold to homeowners as an attractive, natural mulch for their yards and gardens. Unlike the production of turpentine, which eventually weakens the trees beyond recovery, pine straw represents a natural resource that renews itself as new needles replace the old. Some evidence suggests, however, that the removal of the needles reduces the numbers and diversity of arthropods on the forest floor, which, in turn, may lessen the food resources available to ground-foraging wildlife, perhaps including the endangered red-cockaded woodpecker. Moreover, harvesting pine straw also removes the fuel that supports a fire-adapted community and introduces an unnatural disturbance to which these forests and their biota have not adapted.

may magnify these events, and hurricanes and insect attacks may initiate similar settings. Like most pines, longleaf pine is shade intolerant, which means its seedlings cannot prosper in the shade of the parent trees. Hence, longleaf pines do not develop in stands that include trees of different ages and sizes—instead, the trees begin growing in synchrony when the previous stand ends its tenure. Exceptions occur when spotty fires or other local disturbances open up smaller areas in which seedlings can survive, thus initiating a stand of younger trees amid a larger area occupied by an older age class.

Longleaf pine woodlands develop in a variety of soils, but two sites stand out because of their distinctive park-like appearance (i.e., savannas of widely spaced trees). On well-drained sandy soils the understory develops as a lawn of wiregrass (Fig. 6-2, *right*). The species grows in tufts consisting of hundreds of long leaves, most of which are dead, whereas those still alive are fibrous and largely devoid of green cells—a combination that turns the plants into torches ready to ignite. Indeed, wiregrass becomes the primary fuel that carries ground fires across a sandy savanna. Its growing points, however, lie protected just below the soil surface, and, as Washington observed, the plants remain viable and quickly sprout after a fire. Moreover, ground fires during the growing season provide the ecological trigger that stimulates wiregrass to flower and set seed. For the most part, however, wiregrass spreads vegetatively from tillers that expand outward from established clumps invigorated by frequent burning. Wiregrass communities in the southeastern United States include nearly two hundred rare plants, of which sixty-six occur nowhere else. Huckleberry bushes and small hardwoods (e.g., turkey oak) often dot the grassy ground cover, but frequent ground fires check the growth of these and other woody vegetation. Given their easy access, these "dry savannas" were the first to be exploited by commercial interests. Today, the Carolina Sandhills, described later, are among the few areas in North Carolina where significant stands of this type still occur.

Conversely, a "wet savanna" develops where poor drainage produces a water table at or near ground level. As a result, the ground becomes soggy and enriched by an accumulation of organic material, which, coupled with periodic burns, favors a dense ground cover of herbaceous and

FIGURE 6-2. Pine savannas occur in two general types, wet and dry, but both depend on fires to maintain their structure. Wet savannas feature soggy organic soils with a species-rich understory of wildflowers, carnivorous plants, and shrubs (*left*), whereas an understory dominated by highly combustible wiregrass develops on sandy soils in dry savannas (*right*).

woody plants (Fig. 6-2, *left*). One such place is Green Swamp, poorly named but nonetheless an acknowledged hotspot of biodiversity.[2] Small-scale plots in the sixteen-thousand-acre preserve revealed an average of thirty-five species of plants, whereas no more than seventeen species occur in similar plots in woodlands and forests elsewhere in temperate North America. Moreover, plots in the tallgrass prairie, widely known for its diversity, contain an average of eighteen species and never more than twenty-eight species. In all, such data indicate that the diversity

2   Green Swamp, now a preserve protected and managed by The Nature Conservancy, was declared a National Natural Landmark in 1974. Many wet savannas elsewhere in North Carolina, however, were long ago ditched, drained, and converted into farmland.

FIGURE 6-3. Naval stores produced from the resin of longleaf pines were once an important sector in regional economies, including southeastern North Carolina. Today, less than 3 percent of these pine forests remain in the United States. Photo courtesy of New Hanover County Public Library, North Carolina Room.

at wet savannas such as Green Swamp exceeds that at all other North American communities. In assessing these data, ecologists Joan Walker and Robert Peet concluded that the regular influence of fire underlies the remarkable richness of such environments. Indeed, a variety of grasses, sedges, orchids, and other wildflowers flourish at Green Swamp along with fourteen species of carnivorous plants.

Alas, only 3 percent of the estimated ninety million acres of longleaf pine that once covered Bartram's "vast plain" remain today—the result of logging, rooting hogs, urbanization, fire suppression, and preferences for faster-growing trees. Untold numbers of trees were cut outright for their valuable timber, but thousands of others were logged after they were exhausted by years of tapping for naval stores (Fig. 6-3). Logging expanded even farther when steel-hulled steamships, which did not require tar and pitch, lessened the demand for naval stores. Unfortunately, the cut-over tracts were seldom restored with longleaf pine; land managers instead favored faster-growing species, particularly loblolly pine. Historically, lightning strikes initiated ground fires about every two to five years, thereby consuming the fuel on the forest floor before it could accumulate and sustain more intense "hot fires." Indeed, because of its hot and humid summers, the South is particularly prone to spawn thunderstorms well charged with lightning. Given the lack of roads and other anthropogenic firebreaks, such fires once extended over large areas before streams, wetlands, or the lack of fuel at a recently burned location ended their sweep across the landscape. Today, however, networks of roads, supplemented by firefighting organizations, preclude extensive, naturally occurring burns in longleaf pine communities, most of which in fact now depend on a regime of fire management—controlled burning—to survive. Lacking frequent burning, an understory of turkey oaks or other hardwoods develops and inevitably replaces the pines and the rich carpet of herbaceous vegetation (Fig. 6-4).

FIGURE 6-4. Shade-tolerant oaks will replace longleaf pine unless periodic fires sweep across the understory. As shown here, oaks have already gained a foothold in a stand of longleaf pine that has not burned for at least ten years.

Fire Ecology in Brief

Fire occurs as a natural force of nature in virtually all terrestrial communities but varies significantly in its frequency. In their natural state, grasslands burn regularly, whereas fires occur far less often in deserts because the sparse ground litter provides little fuel. On a prairie in North Dakota, for example, lightning ignited as many as twenty-five fires per year. In forests, core borings or cross sections of older trees reveal fire-scarred growth rings that disclose the frequency of fires prior to settlement—often several times per century. Surprisingly, fires rarely kill wildlife, which escape in burrows or simply distance themselves from the flames. For example, ground fires in pine savannas kill few, if any, gopher tortoises, which find refuge in their burrows along with a host of other species. At least one insect, the pitcher plant moth, flees in response to smoke, but the eggs and larval stages of other species may perish in the flames along with other invertebrates. Ground fires may reduce the cover available to amphibians for one to two years, although in the long run fire suppression remains detrimental to these and other herptiles in pine savannas.

Three areas of pines persist within the eastern deciduous forest, a region that otherwise features a climax of hardwoods— various mixes of oaks, maples, beech, and hickories—but only if repeatedly swept by fires. One of these areas includes the range of jack pine, especially around the western Great Lakes. In the wake of regular burning, the regrowth of jack pine within a small area in Michigan provides the sole breeding habitat for Kirtland's warblers, which locate their nests on the ground just under the tips of the lowest branches on the young trees. Lacking fire, however, the jack pines continue growing to the point where their lower branches are too high aboveground to provide the birds with nesting cover. Thanks in large measure to prescribed burning, Kirtland's warblers no longer face extinction and were removed from the federal list of endangered and threatened species in 2019. Fires also maintain the New Jersey Pine Barrens, which features two areas of stunted pitch pines known as pygmy forests; when their seeds are experimentally planted in fire-free sites, they produce only dwarfed trees (i.e., a genetic trait evolving from the selective pressure of fire). Finally, as described more fully in this chapter, fire also maintains longleaf pine–wiregrass communities in the southeastern states.

The pine forests at each of these locations represents a fire climax—a state of development that, without repeated burning, would give way to another type of forest. In each case, the dominant pines have evolved adaptations—the serotinous cones of pitch pine, for example—that facilitate their ability to cope with fires. Indeed, these pines have evolved to the point that fire has become an essential component of their life history.

Fires also protect native grasslands from invasions of woody vegetation, the result of three differences in the two types of vegetation. First, the growing point (buds) of grasses lies at or below ground level and is protected from most fires, whereas the buds of shrubs and trees develop fully exposed at the tips of branches, where they are easily damaged by fire. Second, grasses quickly regrow after a fire and soon regain their full biomass (to grasp this feature, think of how often lawns need mowing), but severely burned woody plants, if they survive, may take years to reach their prefire size. Finally, because grasses regrow so rapidly, they also quickly mature and produce seed, often less than a month after a fire and almost always in the same growing season. By comparison, trees and shrubs require several years or even decades to produce seeds, hence giving grasses a huge reproductive advantage over woody vegetation. Note, however, that cheatgrass, an invasive Eurasian species now widely established on western rangelands, represents a highly flammable fuel that increases the intensity and frequency of wildfires, which may lessen the ability of native grasses to regain a foothold on burned areas.

Controlled (or prescribed) burning offers practical and ecological benefits as a management tool. Defined as the skillful application of fires under exacting conditions (e.g., wind speed) on a specified area for a specific purpose, controlled burning consumes accumulated fuel and thereby lessens the threat of runaway wildfires. For ranchers, controlled burning also replaces unpalatable thatch with tender, protein-rich new growth and retards the encroachment of brush. Similarly, wildlife managers burn marshes to improve habitat for waterfowl and furbearers. Forest ecologists likewise burn blocks of pinelands to keep out hardwoods and improve habitat for fire-dependent plants and wildlife. Singly or in cooperation, federal, state, and private agencies regularly burn locations in North Carolina such as Croatan National Forest, B. W. Wells Savannah, Fort Bragg, Holly Shelter Game Lands, and the Green Swamp Preserve.

Longleaf Pine Specialists: Red-Cockaded Woodpeckers

Red-cockaded woodpeckers (RCWs to insiders) scarcely live up to their name—the cockade, a dash of crimson feathers on the temples of males, is all but invisible under field conditions. (A prominent white check patch on either sex presents a much better field mark.) In 1810, Alexander Wilson named what he thought was a new species he found in "the pine woods of North Carolina." Unbeknownst to him, however, three years earlier the French naturalist Louis Vieillot had described the species that he called "le pic boreal," the northern woodpecker, which grossly mischaracterized the birds' actual distribution; he also presented a somewhat faulty description of its plumage. But to Wilson the small streak of red plumage brought to mind the feathers adorning the head-gear of British soldiers that American colonists had mocked as "maca-roni" in "Yankee Doodle Dandy"—a red cockade. According to the rules of nomenclature, Vieillot's formal usage of *borealis* in the species' Latin name remains intact to this day, whereas Wilson gets credit for the birds' common name, as well as for providing a more accurate description of their appearance.

Instead of being a northern bird, as Vieillot believed (or guessed), RCWs' range extended over much of the southeastern United States from southern New Jersey, south to Florida, and west to eastern Texas. Much of their original distribution coincided with the range of longleaf pine, but as the pines disappeared, so did the birds. In 1970, RCWs appeared on the federal list of endangered species. Of those that remain, most occur on military bases and in national forests; others persist on lands owned by state and private conservation agencies.

Important ecological and behavioral features distinguish RCWs from other woodpeckers, not the least of which is the excavation of their nest-ing cavities in living trees, of which longleaf pine is foremost (Fig. 6-5). In doing so, they utilize the pines' resins to protect their nests from pred-ators, especially any of several subspecies of rat snakes whose respective ranges overlap with the distribution of the birds; all are adept climbers with an appetite for eggs and nestlings. RCWs also form groups consist-ing of up to six birds, a mated pair plus nonbreeding "helpers," usually

FIGURE 6-5. Red-cockaded woodpeckers typically nest in longleaf pines; the birds chip sticky resin pits around the entrance to deter predators. Females, as shown here, lack the red cockade, but even on males this feature is not easily seen.

the male offspring from the previous nesting season. Conversely, young females disperse to establish new breeding groups with males elsewhere. As members of these extended families, the helpers share incubation duties, feeding the nestlings, defending territories, and excavating cavities, the latter being particularly important because of the work required to tunnel in living wood. Each group requires a territory of about two hundred acres of mature longleaf pine.

Only certain trees prove suitable for constructing cavities, most of which serve as roosts for individual birds; a group thus may have up to six cavities in use at one time. Cavity trees must be at least sixty years old, which assures a trunk large enough to physically house a nest, as well as allowing the cavities to be placed high above the heat and smoke from fires on the forest floor. Additionally, older pines are more likely to have red heart fungus, which decays the heartwood and eases the task of excavating a cavity. Still, it takes up to three years and the collective efforts of the group to excavate a cavity. Along the way, the birds also scale away the bark surrounding the entrance hole, leaving an exposed surface, in which they drill resin wells. The resulting flow of resin creates a sticky barrier that effectively deters rat snakes from entering the cavities. Remarkably, the birds can detect those pines that produce especially copious amounts of resin, and of these, the dominant male selects the best producer for the group's nest tree. Likewise, they somehow select pines in which the fungus has had enough time (at least fifteen to twenty years) to soften the heartwood. The occurrence and extent of heartwood decay correlate with the age of the trees. Hence, to be of value to the birds, the pool of suitable cavity trees must include those upward of ninety years old. Regrettably, these are the very trees sought by loggers. In all, when compared with loblolly and other species of pine, only longleaf pines, because of their longevity, fire tolerance, and resilience to storms, provide the optimal characteristics to provide continual nesting and roosting sites for generations of RCWs.

Management efforts to increase suitable habitat include adding artificial nest cavities at sites where pines have been damaged by storms or have not yet grown old enough to have infections of red heart fungus. Initially, biologists drilled cavities, in which resins occasionally accumulated, but they later adopted a technique in which a prefabricated cavity is inserted into a box-shaped recess cut into a tree trunk at least fifteen inches in diameter. Newer designs facilitate inspecting the nests, including access to eggs and nestlings for measurements and banding, as well as clearing out debris from other users (e.g., southern flying squirrels). As hoped, the birds readily accepted the artificial nests, in which nesting success paralleled the results in natural nests (and the trees suffered no permanent damage).

A final note. Because of their protected status, RCWs were not always welcomed by private landowners who feared that the presence of the birds would curtail timber harvests or other activities. The concern was real and thwarted the hope that the private sector might help the birds recover. In response, the U.S. Fish and Wildlife Service initiated Safe

Harbor Agreements in 1995 in North Carolina's Sandhills region, which assured landowners would not suffer additional regulatory actions if they contributed to the recovery of RCWs in ways that increased the population on their lands (e.g., planting longleaf pine seedlings). The program, entirely voluntary, requires that the landowner and agency establish a baseline population estimate. Thereafter, the landowner can make improvements without risking additional regulatory actions provided that the landowner maintains the baseline population. When the agreement expires, landowners must continue to maintain the base population but may undertake activities that might harm a larger population. In short, the program removed disincentives for managing the birds on private lands. Because of their success, Safe Harbor Agreements are now in force for other threatened or endangered species listed under the federal Endangered Species Act. Meanwhile, an assessment has suggested ways to further improve the program's effectiveness.

Big Cones, Big Squirrels

Fox squirrels and longleaf pines occupy essentially the same distributions in North Carolina: the southeastern Coastal Plain and its western fringe of Carolina Sandhills. Their size in this region is larger than that of the subspecies that occur elsewhere in the United States (their range extends across the South and Midwest) and almost twice the size of gray squirrels. In fact, their large size seems associated with their lifestyle. When foraging, they scamper across the open ground between the widely spaced pines instead of traveling tree to tree, as do gray squirrels in deciduous forests packed with oaks, hickories, and other hardwoods. And instead of consuming a diet dominated year-round by acorns, fox squirrels in North Carolina feed heavily on longleaf pine seeds, notably in midsummer before the scrub oaks drop their annual crop of acorns. To do so, however, requires an animal large enough to force open the equally large cones, including those still green and tightly closed. The reward is a meal of large, energy-packed seeds, each of which when mature contains 6.5 kcal per gram. (For comparison, a gram of chocolate provides 5 to 5.75 kcal of energy.)

Old Logs, New Floors

From the earliest days of settlement, the original pine forests and cypress swamps on the Coastal Plain attracted loggers. The trees were huge and there for the taking by doughty souls equipped with axes, saws, and plenty of muscle. As might be expected, the stands near major rivers, of which the Cape Fear was foremost, were the first to fall; the river and its larger tributaries provided cheap transportation to mills downstream. Business was good—no fewer than forty lumber mills operated along the Cape Fear in 1764. River drivers lashed the logs into rafts, some of which negotiated whitewater rapids before reaching the tranquil waters below the Fall Line. Whether the river was calm or not, some logs broke loose and sank, especially at what were once holding pens near long-gone mills, sites now mother lodes for "sinkers." Some estimates suggest that up to 35 percent of the logs floated downstream during the 1700s and 1800s did not survive the voyage. Ironically, the wood in these old logs, because of its densely packed growth rings and greater weight, was the best of its kind, making it especially valuable—if it could be retrieved.

While it seems counterintuitive, sunken logs undergo little change. Once on the bottom, the logs lie in an oxygen-depleted environment where decomposition proceeds almost imperceptibly, and those buried in sediments decay even more slowly. The wood not only retains its structural integrity but, when freshly sawn, still emits a woody scent.

Moreover, the long years underwater actually improve the wood's natural luster and appeal. But centuries would pass before suitable equipment and technology could detect and resurrect the logs from their watery crypts.

Then, in the 1990s, a new kind of logger arrived to claim these sunken treasures. Once a trove of logs is located, sometimes with divers but more often with the aid of sonar equipment, a crane and grapple haul the prizes aboard a barge, after which they renew their long-delayed journey to a sawmill. Authorization to remove the logs required that trees with roots and branches be returned intact to the riverbed, thus maintaining the natural occurrence of such debris on the bottom. Additionally, antiquities discovered during the logging operations are turned over to the state's Underwater Archaeology Branch.

Because the old-growth trees were up to three hundred years old when logged, the wood may be as much as five hundred years old when retrieved. Most of the wood—attractively tinged by tannins in the water—ends up as flooring in upper-end dwellings but also as fine furniture, hand-hewn beams, and even musical instruments. Boards from a single cypress log may command $10,000, which seems certain to assure that this modern form of logging will continue. And according to experts on the subject, thousands of logs still remain underwater, where they patiently await their destiny as ancient sources of modern products.

FIGURE 6-6. The pelage of fox squirrels varies greatly, ranging from entirely black to those with black heads, white noses, and gray bodies often tinged with rusty orange—variations that characterize populations living in southern pine forests subject to frequent fires. Elsewhere in their extensive range these large squirrels bear reddish-gray pelage with little or no black coloration.

For fox squirrels, no feature is more prominent than their "coats of many colors."[3] Individuals in the Midwest, for example, have reddish-gray body pelage, whereas some gray individuals elsewhere have a black head with a white nose, as well as black toes; others have black dorsal pelage, and still others are entirely black (Fig. 6-6). Notably, the darker morphs occur most often in the Coastal Plain of North Carolina and other southeastern states; hence, the trait may be associated with the regular occurrence of fires in longleaf pine forests. A positive correlation in fact exists between the incidence of lightning-initiated fires and the occurrence of melanism in fox squirrels on the Coastal Plain. Such a scenario suggests that these individuals might gain the advantage of protective coloration when foraging in frequently burned landscapes. Stated otherwise, the darker pelage may be an adaptation that lessens the risk of attacks by aerial predators. Indeed, experimental evidence indicates that red-tailed hawks, known predators of fox squirrels, respond more slowly to moving models of squirrels with intermediate amounts of black

3　While collaborating with John James Audubon (1785–1851) on *The Viviparous Quadrupeds of North America*, Lutheran minister John Bachman (1790–1874) advised the famed painter that "the ever-varying squirrels seem sent by Satan himself to puzzle the naturalists." Indeed, for many years naturalists struggled with the taxonomy of fox squirrels, which are now recognized as a single species with ten subspecies. Bachman's sparrow, a resident of southeastern pine forests, and Bachman's warbler, now considered extinct, are among the species named in his honor. Bachman also discovered the marsh rice rat.

dorsal pelage in comparison with other color patterns. This and other tests suggest that fox squirrels with such coloration likely improve their survival and thus would favor the persistence of the responsible genes in fox squirrel populations on the Coastal Plain. In comparison, note that the preferred habitat—hardwood forests—of gray squirrels burns on average perhaps twice per century and that melanistic pelage occurs much less frequently in their populations.

In addition to pine seeds and acorns, fox squirrels eat lesser amounts of other foods, including fungi that mature underground. After excavating and eating the fungi, the squirrels passively disperse the indigestible spores throughout the pine forest in their droppings, much to the advantage of the fungi's reproductive success. Moreover, the thread-like filaments ("mycelia") of the fungi interact with the tree roots in a mycorrhizal relationship: the fungi, incapable of photosynthesis, receive carbohydrates from the trees, and the pines gain more efficient access to water and nutrients in the soil. In all, a picture emerges of an integrated system involving fox squirrels, fungi, and longleaf pines.

Venomous and Vulnerable

Eastern diamondback rattlesnakes reach their northern limit in southeastern North Carolina, from there extending south along the Coastal Plain to Florida and the Gulf Coast to eastern Louisiana. Overall, the original range of the species largely coincided with the historic distribution of longleaf pine, but the extensive loss of longleaf pine forests likewise severely contracted the distribution of these snakes. (Diamondbacks, however, do occur in other types of habitat, varying from marsh edges to scrublands.) Perhaps to the chagrin of Texans, these are the largest venomous snakes in North America, reaching maximum lengths of eight feet, whereas the largest western diamondback on record is a full foot shorter. However, most individuals of either species range between three and six feet in length. Eastern diamondbacks bear a distinctive pattern of dark diamond-shaped markings with cream-colored borders; a black band extending from their eyes, also with cream-colored edges, further identifies the species (Fig. 6-7).

FIGURE 6-7. Although not inherently aggressive, eastern diamondback rattlesnakes nonetheless will strike when threatened and, as shown here, adopt a posture with an extensive forward reach. Because of extensive habitat loss and human persecution and exploitation, the species appears on North Carolina's list of endangered species.

Their distribution in North Carolina covers the lower southeastern corner of the Coastal Plain but does not extend to the Sandhills Region along the Fall Line (see below). Pine savannas offer prime habitat for eastern diamondbacks, although they also occur in overgrown fields near forests and drier pocosins. Eastern diamondbacks feed on small rodents, rabbits, and occasionally birds, commonly remaining coiled and motionless for extended periods of time waiting for passing prey.

In much of their range, they seek shelter and nest sites in tunnels excavated by gopher tortoises, but lacking these in North Carolina, they instead utilize hollow logs, pockets in the base of dead trees, and stump holes; in winter, individuals also hibernate alone in these sites. Males battle each other when competing for territories and females in a "combat dance," in which they elevate and entwine their bodies, then thrash about until one retreats—a display that was once believed to be court-ship and mating behavior between a male and female. Females incubate and hatch their eggs internally, hence giving birth to live young, a process formally known as ovoviviparous reproduction; clutch size varies considerably but averages about fifteen. The hatchlings bear markings similar to those of the adults and enter the world fully equipped to deliver a venomous bite.

In proportion to their body size, the fangs of eastern diamondbacks are the largest of the more than thirty species of rattlesnakes and deliver a potent venom—a hemotoxin—that kills red blood cells and damages tissues. Larger individuals may have one-inch fangs and deliver a high yield of venom, a combination that makes these rattlesnakes the most dangerous venomous snake in North America. Fortunately, they seldom attack without cause, but when provoked they can strike outward for one-third or more of their body length.

Humans have long exploited eastern diamondback rattlesnakes, in the past including carnival-like "rattlesnake roundups" that inhumanely obtained the snakes by gassing their dens. Because of their attractive markings, diamondbacks are sometimes hunted for the commercial value of their skins, and, intentionally or not, others become roadkills; still others are shot as vermin. Such direct causes of mortality, coupled with the widespread loss of longleaf pine savannas, have severely reduced populations of these snakes in many areas. At present, however, only North Carolina fully protects the species, and federal action on a petition to have it formally listed as threatened remains stalled. While some see only the fearsome aspect of the species, most naturalists view the eastern diamondback rattlesnake as a much persecuted and unique cog in nature's machinery that warrants greater protection lest the species slips further downward.

Carnivorous Plants

Sites with wet, acidic, and nutrient-deficient soils present difficulties for most plants, but not for those that survive on nutrients—primarily nitrogen and phosphorous—gained from insects, spiders, and other small creatures they capture and digest. They do, however, rely on photosynthesis in the same fashion as other green plants to produce energy. But the unique structure and habits of carnivorous plants evolved at a cost—for the most part, they cannot compete with other plants. As a result, carnivorous plants thrive only where poor growing conditions keep out most would-be competitors. In other words, they are specialists with a narrow niche.

The carnivorous plants most familiar to naturalists come in three basic models. Some employ active mechanisms akin to a snap trap to capture their prey, whereas others function passively as a pitfall or as sticky flypaper. These include, respectively, the Venus flytrap, pitcher plants, and sundews (Fig. 6-8). Bladderworts, a large genus of mostly submersed species, display another type of mechanism: they trap their prey inside small vacuum-filled bladders that suddenly open and suck in prey such as tiny aquatic crustaceans known as water fleas. The process, which might be likened to a "gulp," begins when the victim touches a trigger hair, which opens the bladder and creates an incoming rush of water containing the prey; the opening quickly closes, entrapping the hapless victim. The entire process takes less than fifteen milliseconds. Butterworts have evolved yet another means of capturing and digesting prey: they curl the edges of their leaves around victims mired on the sticky ("buttery") foliage. Three species occur in North Carolina, of which the blueflower butterwort is often found in the Carolina Sandhills (see below). Because of their varied diet, all of

FIGURE 6-8. Carnivorous plants of North Carolina's Coastal Plain include, *top to bottom*, species such as pink sundews, yellow pitcher plants, and Venus flytraps, each with unique adaptations for capturing prey.

these plants—flytraps to butterworts—are better described as carnivorous rather than insectivorous. Either way, the words of North Carolina's pioneering ecologist B. W. Wells capture the moment: "Nature has carried out the most amazing different modifications of leaf parts to arrive at the same end—the capture and ingestion of insects."

Because of their uniqueness, Venus flytraps head the list of interesting carnivorous plants occurring in North Carolina. Charles Darwin in fact lauded the Venus flytrap as "the most wonderful plant in the world," praise indeed from the founding father of evolutionary biology. The species, the sole member of its genus and found only within a sixty-five-mile radius of Wilmington, is one of only two species of more than five hundred other carnivorous plants worldwide to develop a snap-trap mechanism for capturing prey. (The other, the waterwheel plant, grows underwater in tropical areas of Australia, Asia, and Africa.) As such, it was one of the most popular items sought by John Bartram's British patrons. Credit for discovering the species, however, goes to Arthur Dobbs, who served as the royal governor of North Carolina between 1754 and 1765. (Dobbs is also credited with discovering the role of insects as pollinators.) "The most wonderful plant" became the official state carnivorous plant of North Carolina in 2005.

The "trap" is actually the tip of a highly modified leaf that opens and closes in the manner of a clamshell. A row of bristles projects from the outer edge of each lobe; the bristles mesh together like interlocking fingers when the trap closes. The remainder of the leaf—a flattened petiole—provides a surface for photosynthetic activity. The leaves arise from a bulb-like underground structure, as does a stem about four inches high that in season bears a cluster of white flowers.

When lured into the open trap by a sweet-smelling nectar, insects or other prey come into contact with hairs known as trichomes. If, after touching one of these "triggers," the prey again touches another within twenty seconds, the trap snaps shut—a reaction taking less than one-tenth of a second. Five more touches initiate the secretion of enzymes and the beginning of digestion. Lacking a second touch, however, the trap remains open, which negates a fruitless response from a raindrop or a falling leaf. Remarkably, these reactions indicate that Venus flytraps have a sense of memory that includes counting. Moreover, cells on the inner surface of the trap provide several functions: nectar and enzyme production and the absorption of nutrients.

The menu for Venus flytraps varies somewhat between locations, but in general ants form about a third of the diet, closely followed by spiders and lesser amounts of beetles and grasshoppers; only a few flying insects become victims. Smaller prey, which offer too little return for the effort, can escape through the narrow spaces between the bristles, but the struggles of a larger victim close the trap ever tighter. Overall, the traps capture about one in four visitors—seemingly low but actually better than the success rate of most animal predators (e.g., wolves). After a week or so of digestion, the trap reopens and releases the exoskeleton, now no more than an empty husk. A trap remains functional for up to five meals, then takes part in photosynthesis for two to three months before the entire leaf drops off.

Unfortunately, poachers at times have removed thousands of Venus flytraps from natural areas despite the ease with which the plants can be legally and rather easily propagated for sale. However, in 2014, the crime was elevated from a misdemeanor with a fifty-dollar fine to a felony that can send a poacher to prison for twenty-five to thirty-nine months.

Pitcher plants, exemplars of the pitfall strategy, also attract their prey with alluring scents but otherwise react passively to the arrival of their victims. Ants again top the list of foods, but many other arthropods also provide pitcher plants with nourishment; on rare occasions, even small vertebrates may be captured. Once inside the pitcher—functionally a funnel—the prey pass by curved hairs that allow further movements deeper into the narrowing passage but hinder any efforts to reverse course and escape. Still, the capture rate falls below 1 percent, which may represent a means of precluding prey from developing avoidance behavior that would even further decrease the success rate (i.e., so many prey survive that they do not evolve a specific defensive mechanism via natural selection). Eventually, the trapped prey drown in the pool of water inside the plant, which in some species is enriched with enzymes. As the dead organisms break down, they supply the plant with nutrients.

Yellow pitcher plants occur almost exclusively on the Coastal Plain, especially in wet pine savannas. These showy plants feature wide overhanging hoods and flowers that emerge before their leaves develop, likely an adaptation that prevents their own pollinating insects from becoming victims. The pool of water inside yellow pitcher plants contains digestive enzymes. Two subspecies of purple pitcher plants occur in North Carolina: a rare form in the mountains, and the other more widely distributed across the Coastal Plain. Both forms rely on bacteria, not enzymes, to break down their prey. These and the two other species of pitcher plants are protected in North Carolina.

Still, there's more to the hidden world inside a pitcher plant, where a miniature aquatic community goes about its business, enzymes or not. These unique environments, formally known as phytotelmata, provide habitat for, among others, protozoans, rotifers, crustaceans, insects, and, in some locations, amphibians; animals living in phytotelmata are collectively known as inquilines. The phytotelmata of purple pitcher plants, for example, host the larvae of three species of flies, each eating detritus but doing so at different locations in the water column, top, middle, and bottom, thereby avoiding competition via an ecological strategy called resource partitioning.

Three species of pitcher plant moths also depend on their namesakes, two of which rely solely on a single species of pitcher plant to complete their life cycles—one each, respectively, on purple and yellow pitcher plants. The third is a generalist lacking host specificity—any pitcher plant will meet its needs. Females of the latter lay one to three eggs inside the pitcher, and after hatching, the larvae feed on the plant's tissues, at times causing a wilted leaf. Small projections on the sides of the larvae catch on the hairs inside the pitcher, impeding falls into the water below; the larvae also produce strands of silk that serve as safety lines, on which they can, if needed, climb back to safety. The larvae eventually cut a small drainage hole through the pitcher, above which they construct a silk pupal chamber on the inside of the pitcher that, thanks to their engineering, will remain dry even if the water level rises. After emerging from their pupae, the half-inch adult moths—distinguished by broad black and straw-colored bands—have evolved the unique ability to walk safely through the hairs inside the pitcher that prevent the escape of other insects. They remain stationary inside the pitcher during the day; at night, they feed on nectar at the upper edge of the pitcher, from which they also launch flights in search of mates.

Other species associate with pitcher plants but not in an obligatory relationship (i.e., they frequent other vegetation as well and hence display a facultative association with pitcher plants). One of these, the green lynx spider, builds no web but instead pounces cat-like on prey as they enter the pitcher, literally stealing prey from the plant. To facilitate the ambush, the color of these spiders closely matches that of the plant. The diet of green lynx spiders includes almost all kinds of the arthropods they encounter, although their venom-injecting "jaws" (chelicerae) cannot pierce the hardened outer wings (elytra) of beetles.

Sundews get their name from the dew-like sparkle of the sticky drops that tip the tentacles ("hairs") arising from their leaves—they glisten in sunlight but in reality entrap gnats, mosquitoes, and other small insects attracted by the adhesive's fragrance. As a victim struggles, the flexible tentacle moves toward other tentacles and together wrap the prey in a fatal grasp; enzymes then begin the digestive process.

Four of the five species of sundews in North Carolina occur in the Coastal Plain. (The fifth favors mountain bogs.) Of these, pink sundews, the most common, sometimes cover wet ground with a carpet of rosettes no larger than a fifty-cent coin. Spoonleaf sundews prefer even wetter sites such as a water-filled depression or rut. Accordingly, their long leaves rise above the standing water to trap insects. In contrast, the tiny rosette of the dwarf sundew, about the diameter of a hefty pencil, may be overlooked; its showy white flower is about the same size. Besides their sticky leaves, hairs elsewhere on these plants may also trap insects. Threadleaf sundews lack the broad leaves of the other species and instead bear thin, foot-long leaves with a furry cover of hairs that may enable the capture of larger insects. The species is rare and confined to the southeastern corner of North Carolina.

The interplay between pink sundews and funnel-web wolf spiders because of their overlapping diets suggests an example of competition between highly dissimilar species, in fact, those of different kingdoms.

Under field conditions where the spiders and plants shared a large area but in different densities, the spiders constructed their webs away from the sundews and increased the area of their webs—results that argue for competition for prey between the two species. Moreover, laboratory experiments revealed that the spiders could capture enough prey to lessen the nutrition available to the sundews to the point where the plants produced fewer flowers (by half) and seeds. In evolutionary terms, these results suggest that the spiders could reduce sundews' "fitness" by negatively affecting their ability to reproduce. All told, whether it be the curious interior of pitcher plants, the trigger itch of Venus flytraps, or spiders vying with sundews, carnivorous plants in their own way capture naturalists, as well as insects.

Inland Dunes: The Carolina Sandhills

A rolling terrain known as the Carolina Sandhills covers three million acres of the Coastal Plain more than one hundred miles inland from the Atlantic coastline, much of it in Scotland, Hoke, Richmond, and Moore Counties. The sandy hills in North Carolina lie at the northern end of a nine-to-thirty-seven-mile-wide ribbon of similar landscape that extends southward across South Carolina and into Georgia (Fig. 6-9). In the Carolinas, the region lies between the Fall Line and the Orangeburg Scarp, a shoreline that marks the limit of the ocean during the Pliocene some three million years ago.

In the past, geologists proposed that the sands originated as beach deposits or were carried downstream by ancient rivers. Newer work indicates that the sands originated from an older, underlying bed of Cretaceous sandstones known as the Middendorf Formation. The resulting blanket, known as the Pinehurst Formation, consists of dunes up to twenty feet tall and sand sheets that vary from about two to thirty feet in depth. The latter developed in several episodes, each highlighted by cooler temperatures and reduced plant cover; the first of these events began some seventy-five thousand years ago, and the last ended about six thousand years ago. Since then, the climate has warmed, and long-leaf pine savannas now stabilize much of the sandy terrain. Elsewhere,

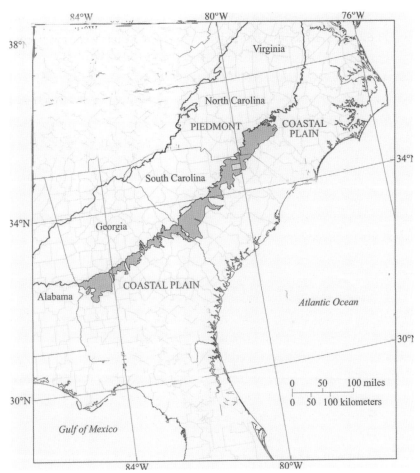

FIGURE 6-9. The Carolina Sandhills extend into North Carolina, where they form the southwestern edge of the Coastal Plain. Map courtesy of the U.S. Geological Survey.

however, local conditions—some wet, some extremely arid—establish other communities.

The two geologic formations also interact to influence the development of some local communities not otherwise typical of a sandy region. Because rainwater moving downward through the porous sands moves laterally when it encounters the nearly impervious Middendorf Formation, it emerges in places as seeps and steamheads where wetland plants can survive. The communities arising at these sites—patches of pitcher plants, for example—add to the exceptional diversity of Carolina Sandhills flora. For animals, the rapid absorption of surface water means they must respond quickly to its availability. Some, particularly those amphibians that lay their eggs in water, necessarily breed quickly when rains provide pools of surface water. The water-dependent larval stages, specifically tadpoles, of these species also must develop rapidly into adults before the pools dry, and these species have indeed evolved a quickened pace of maturation to accomplish this feat (see below).

Vast forests of longleaf pine once covered the Sandhills, but, as they have elsewhere in the Southeast, lumbering, farming, and urban development have vastly reduced the acreage to a fraction of its former extent. Today the only significant tracts of the region's pine forests remain at protected locations such as Fort Bragg, Weymouth Woods Sandhills Nature Preserve, and, in South Carolina, the Carolina Sandhills National Wildlife Refuge. When burned regularly, these stands develop as savannas, as shown in Figure 6-2. Small areas of bare sand in the region support turkey oaks, so named because the shape of their leaves commonly resembles a turkey's foot, and scattered tufts of wiregrass (Fig. 6-10).

The ground cover in the Carolina Sandhills includes species with limited distributions, including a few found nowhere else (Fig. 6-11). One of these, littleleaf pixie-moss, is known from just six counties in North and South Carolina; within this range, the diminutive plant occurs with greater frequency at Fort Bragg, but just why is unclear, as seemingly good habitat can be found elsewhere in the Sandhills region. One of just two species in its genus, littleleaf pixie-moss was discovered in 1927 by ecologist B. W. Wells; hence, some botanists refer to the species as Wells' pixie-moss. These compact plants grow low to the ground and appear moss-like when not in bloom; in season, they produce tiny white flowers barely five millimeters wide and resemble patches of snow. They bloom early, commonly in February, and, like several other plants in the Sandhills, have adapted to fire. Another species, white wicky, also occurs in a handful of adjacent sandhill counties in North and South Carolina. In summer, it bears clusters of white flowers, each petal highlighted by faint red streaks. Unlike other shrubs in the genus *Kalmia*, white wicky

FIGURE 6-10. Some sites in the Sandhills with exposed sand support little more than scraggly oaks and scattered clumps of wiregrass, whereas more typical habitat for the region appears in the backdrop of longleaf pine, as shown here at the Carolina Sandhills National Wildlife Refuge in South Carolina. Photo credit: Kay McCrutcheon.

FIGURE 6-11. Plants associated with the Carolina Sandhills include (*top to bottom*) the endangered Michaux's sumac, white wicky, and littleleaf pixie-moss. Photo credit: Alicia G. Jackson (*middle*).

sheds its foliage each year, but before falling, the leaves turn scarlet—a brilliant sign of the changing seasons.

The Sandhills region is the stronghold for Michaux's sumac, which appears on the federal list of endangered species.[4] A dense, hairy shrub of three feet, it favors sandy soils on disturbed sites, especially in those areas with periodic fires; hence, fire suppression has greatly precluded its presence in many areas where it once occurred. Moreover, the species does not have a high reproductive potential in the wild, so propagation becomes a viable management option for its recovery. Fortunately, populations of Michaux's sumac occur at Fort Bragg and other military bases, where they are protected and regularly monitored. Roughleaf loosestrife, another endangered species, grows on moist sites, often in association with pitcher plants (see above). Fires stimulate its reproductive activities by initiating the growth of a terminal raceme of yellow flowers, although its long rhizomes also enable the formation of colonies. Roughleaf loosestrife occurs only in the Carolinas.

Small headwater streams, running fast and cool, in the Carolina Sandhills provide habitat for two endemic fishes (Fig. 6-12). Aptly named, the sandhills chub occurs in shallow streams with sand and gravel bottoms in the drainages of the Lumbee, Cape Fear, and Pee Dee Rivers. Notably, these robust-bodied chubs, which reach lengths of nearly ten inches, prefer sites lacking aquatic vegetation. The sides of the pinkish males feature an olive-green stripe along with red-orange ventral fins, but the drab females and juveniles lack these colorful markings. Sandhill chubs lack the characteristic black spot on the dorsal fin that distinguishes the related and more widely occurring creek chub.

When breeding, male sandhills chubs excavate pits at the tail end of pools, in which females deposit their eggs. When fertilization begins, the males remove and relocate pebbles from the downstream edge of the pit to its upstream end, thereby covering the fertilized eggs. This process,

4   The explorer-botanist André Michaux (1746–1802) traveled widely on behalf of the French government in search of plants to improve agriculture, medicine, and forestry in his homeland. In doing so, he named 283 species collected in the Carolinas, including the sumac renamed in his honor and wiregrass. Another species bearing his name, the Carolina lily, is the official state wildflower of North Carolina.

FIGURE 6-12. Pinewoods darter (*top*) and sandhills chub (*bottom*), both endemic to the Carolina Sandhills, require protection and close monitoring.

which continues as spawning progresses, fills in the pit with a ridge of deposited material. Hence, the ridge grows ever longer upstream, while the pit continually moves downstream (i.e., not unlike a worker digging a trench, inserting a length of pipe, and covering it behind him as he moves along). The size of these "pit-ridge nests" varies with the size of the males and the duration of spawning; some ridges may be nearly four feet long and seven inches high and cover pits about three inches deep. The structure of the ridges promotes eddies that retain the nonadhesive eggs within the confines of the nests. Males thereafter remain on guard at the site for extended periods of time.

The pinewoods darter is narrowly confined in separate populations in the Carolina Sandhills, one each in the respective upper reaches of the Lumber and Little Pee Dee Rivers. Because of their separation, the two populations have developed with distinctive genetic signatures, each forming what are known as evolutionarily significant units. This indicates a long period of isolation for each of the two populations and suggests that the genetic integrity of each deserves the attention of conservation biologists. Like other darters, pinewoods darters have two dorsal fins, the first of which in this species features a bright-red edge. The lateral line is distinctively yellow, and black spots dot the lower head and breast. They range from two to three inches in length. Unlike the sandhills chub, pinewoods darters prefer sites where sunlight facilitates the growth of aquatic vegetation. Both species, however, share a similar diet consisting primarily of aquatic insect larvae.

Several interesting amphibians live in the Sandhills, although not exclusively (Fig. 6-13). One, the Carolina gopher frog, resembles a toad because of its warty skin, large head, and chunky body. Individuals range in color from pale gray to tan, and some are nearly black, but all are spotted; when exposed, the skin folds reveal their yellow-orange coloration. Gopher frogs, listed in 2017 by North Carolina as endangered, once thrived in at least twenty-three populations in North Carolina but today hang on in only seven, including two in the Carolina Sandhills (e.g., at Fort Bragg). In addition to monitoring the remaining populations, a proactive strategy known as head-starting strives to overcome the huge losses of eggs and tadpoles in the wild. To accomplish this goal, biologists rear the tadpoles in captivity; after metamorphosis, the juvenile frogs are released at the same location where the eggs were collected.

In much of their range, gopher frogs are closely associated with gopher tortoises. In North Carolina, however, gopher frog populations lie well north of the range of the tortoises. Where the ranges of these species overlap, the burrows dug by the tortoises provide shelter for the frogs—hence their name—and a considerable number of other species as well, including snakes and small mammals. In North Carolina, gopher frogs instead burrow in root tunnels and stump holes created by decaying trees. Additionally, the holes excavated by crawfish and mammals also serve as burrows for gopher frogs (which were known in the past as crawfish frogs, but the latter are now regarded as a separate but related species). Adults usually live in separate burrows, which they use from year to year. They hunt at night from perches cleared at the burrow's entrance, from which they ambush passing beetles, crickets, spiders, and other invertebrates. Otherwise, the secretive gopher frogs seldom leave

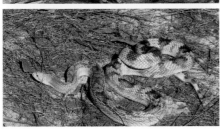

FIGURE 6-13. The fauna of the Carolina Sandhills includes (*top to bottom*) Pine Barrens treefrogs, Carolina gopher frogs, and northern pine snakes. Photo credit: William C. Alexander (*middle, bottom*).

their burrows except to breed, which takes place in temporary ponds (i.e., those that dry each year and thus lack fish populations). The ponds and burrows may be more than two miles apart, which calls for a lot of serious hopping. Curiously, if not humorously, gopher frogs assume a "peek-a-boo" posture when threatened by covering their faces with their front feet.

Another species, the eastern spadefoot toad, derives its name from a hardened black, sharp-edged "spade" projecting from the inner edge of their rear feet. This feature, the hallmark of the group, enables the toads to dig vertical burrows in sandy soils with a slight rocking motion. In dry periods, spadefoots may enter a torpid state and remain underground for several weeks, during which they survive despite losing water in amounts that may reduce their body weight by 40 percent or more. As an additional distinction, the eyes of spadefoots in full light display vertical, cat-like pupils, whereas they are elliptical in other groups of toads.

Eastern spadefoot toads adapt to the irregular availability of rainwater with an unusually short period of maturation—egg to tadpole to adult—which can be completed in less than thirty days (i.e., before the surface water disappears). They share this ability with others of their genus that live in the arid regions of the western United States where rainfall is also unpredictable. This adaptation allows spadefoot toads to breed in shallow pools without interacting with other species of amphibians whose metamorphosis proceeds at a slower pace—in short, they gain exclusive "rights" to temporary but crucial breeding habitat. This ability is of particular significance in the southeastern United States, where the region's diverse amphibian fauna typically results in competition and predation among the various species of tadpoles sharing the same habitat. Moreover, spadefoots quickly breed in response to heavy rainfall during the warm months of the year, which often results in the sudden appearance of an exploding population. Conversely, they may altogether skip breeding in years with unsuitable conditions.

Many herpetologists consider the Pine Barrens treefrog as the most attractive species of its kind in North America. A distinctive purple stripe runs along each side of the green body, and the front feet look like "gloves" of the same color; hidden surfaces on the legs, when exposed, reveal numerous small yellow spots set against a bright orange background. The species occurs in widely separated populations, which include the Pine Barrens of New Jersey, the Florida Panhandle (including a small adjacent area in Alabama), and all of the sandhills of North Carolina and an adjacent part of South Carolina. Because development has claimed much of their prime habitat (pine savannas), Pine Barrens treefrogs have been accorded various levels of protection in each of the states where they occur. North Carolina went one step further when it recognized the species as the official state frog in 2013.

These colorful but secretive treefrogs live in wetlands throughout the Carolina Sandhills, where the acidic water apparently provides the species with a measure of exclusivity. Throughout their range they thrive in waters with pH values of 3.8 to 5.0. However, laboratory experiments indicate that the tadpoles of potential competitors can indeed tolerate the same degree of acidity as that tolerated by Pine Barrens treefrogs, although the eggs of these species were not similarly tested. For now, the reason for the exclusion of other amphibians, especially predators, from the breeding habitat of Pine Barren treefrogs remains uncertain.

In spring, the males attract females with a honking "quonk-quonk, quonk-quonk," which they produce in bursts by inflating a balloon-like pouch on their throats. A chorus of males calling from a single location is not uncommon. When mating, the male clasps the female from behind with his front legs—a sexual embrace known as amplexus—and fertilizes each of five hundred to one thousand eggs, laid one at a time. Depending on water temperature, the eggs hatch in seven to fourteen days, producing tadpoles, which emerge from the water as froglets eighty to one hundred days later. Males reach breeding age in two years, whereas females require an additional year to reach sexual maturity.

The distribution of northern pine snakes includes isolated populations, one of which occurs in the Carolina Sandhills and another in the New Jersey Pine Barrens, where the species has been extensively studied. Powerful constrictors with a diet favoring small mammals, these handsome snakes usually grow to lengths of four to five feet, although some exceed six feet. Large shield-like scales on their snouts facilitate digging winter hibernacula and, for females, nesting sites. In spring, after shedding their worn skins, females leave scent trails of pheromones that attract males. About a month after mating, the gravid females seek sunny, sparsely vegetated locations where the sandy soil is suitable for digging and a fibrous root system of grasses will prevent cave-ins. They excavate tunnels up to five feet long that expand into nest chambers, in which they lay five to twelve eggs. The chambers lie just six to seven inches below the surface, which allows solar heat to incubate the eggs. Telltale mounds of sand mark the entrance to these excavations. At times, other females share a tunnel, but each digs a separate egg chamber. After they have completed their clutches, female pine snakes abandon the site, after which foxes, raccoons, and other egg predators may destroy as many as 25 percent of the nests. North Carolina lists northern pine snakes as a threatened species.

Other upland sites on the Coastal Plain feature large areas of exposed sand, notably in the southeastern corner of North Carolina (e.g., Cape Fear Peninsula and Bladen Lakes State Forest); the coarse, sugar-like white sand bears witness to the ocean that once covered the region. The aridity at these locations, often known as the Sand Barrens, originates not from the lack of precipitation but from the rapid drainage of surface water and the heat generated by the unimpeded radiation—downward as well as reflected upward from the brilliant sand—of the sun's energy. Scraggly turkey oaks dominate the vegetation (Fig. 6-14). As seedlings, the oaks deal with the radiation by turning their leaves vertically (i.e., perpendicular to the ground). B. W. Wells cleverly demonstrated this adaptation when he restricted the movement of the leaves, which resulted in heat-related damage to their tissues. Widely scattered longleaf pines, most small in stature, form what little canopy exists in the Sand Barrens.

For the most part, the extreme temperatures and aridity in the Sand Barrens limit the development and diversity of herbaceous ground cover. Nonetheless, three plants that have adapted dissimilar ways to deal with local conditions include Carolina sandwort, or simply "sand plant," whose fleshy roots store enough water to enable the plants to survive the arid conditions. Pineland scalypink copes in a different way; its leaves form in the fall but fade away by the end of winter, hence eliminating surfaces subject to evapotranspiration during the heat of summer. Likewise, its thin, wire-like stems present a minimum amount of surface area from which precious water can escape into the atmosphere. Bull nettles offer a third strategy: they conserve water by protecting their tissues from injury from would-be herbivores. True to their name, bull nettles—leaves and stems alike—bear hollow hairs that arise from glands filled with an acrid fluid. When touched, the hairs break off and deliver a burning sting. As might be expected, plants of the southwestern deserts have evolved similar adaptations for conserving water.

Seemingly out of place on a sandy landscape, cushions of lichens nonetheless dot the surface of the nutrient-starved soils. Lichens, famous for their symbiotic association of algae and fungi (and sometimes including cyanobacteria as well), occur worldwide in a broad range of environments from the Arctic to the tropics. In appearance, lichens sort out into three groups: foliose (leaf-like clusters), crustose (like paint splashes), and fruticose (branched tangles). Of these, a fruticose form known as dixie reindeer lichen ("moss") is among the lichens that cope with the harsh environment of the Sand Barrens; not being rooted, a cushion can be moved across the sand with a flick of a toe. No reindeer, of course, occur anywhere in North Carolina, but this lichen closely resembles another species that indeed provides an important winter food for caribou, the North American counterpart of Eurasian reindeer. Like lichens elsewhere, dixie reindeer lichens fare well with aridity but not with fire, and their existence remains limited to sites lacking sufficient fuel to carry a fire. Mixed among the lichens are dark, low-lying clumps of sand spikemoss, which, unlike the lichens, are vascular plants with shallow, thread-like root systems. Given the lack of combustible litter, the Sand Barrens community burns only irregularly, and its maintenance is not fire dependent.

Bombs, Beavers, and Butterflies

St. Francis' satyr may be the rarest butterfly in North America. Discovered in 1983, the species then, as now, survives solely in a small population at Fort Bragg Military Installation. The brown wings of this small but attractive butterfly feature a border of dark "eye spots" with yellow edging and flecks of white (Fig. 6-15). Adults do not feed on nectar or any other food during their three-to-four-day life span. As larvae, however, the caterpillars apparently forage only on Mitchell's sedge, although further study of their natural history may reveal a diet that includes other sedges as well.[5]

---

5   The Latin names for both the butterfly and sedge honor Elisha Mitchell (1793–1857). Widely regarded as North Carolina's foremost geologist and naturalist, Mitchell established that a peak in the Black Mountains near Asheville reigned as the highest mountain east of the Mississippi River. His claim was challenged but was later confirmed by the U.S. Geological Survey. Ironically, he fell to his death while rechecking his original measurements; his tomb today sits atop what is now known as Mount Mitchell (elevation 6,684 feet above sea level).

FIGURE 6-14. Brilliant white sand provides the only soil in large areas of the southeastern Coastal Plain, as shown here at Bladen Lakes State Forest. Turkey oaks and a sparse ground cover of lichens (light-colored cushions) and spiny spikemoss (dark clumps) eke out a stressful existence at these sites; low soil moisture and heat reflected by the sand curtail development of most herbaceous plants. Photo credit: Hans-Christian Rohr, North Carolina Forest Service.

FIGURE 6-15. One of the world's rarest butterflies, Saint Francis' satyr survives only in a small area of the Carolina Sandhills in North Carolina. Photo credit: Becky Harrison, U.S. Fish and Wildlife Service.

The population at Fort Bragg, already scarce when discovered, continued its decline and seemed to have fallen into the abyss of extinction. In 1992, however, the butterflies were "rediscovered" in the same unlikely environment: the impact zone on an artillery range. By any measure, the habitat there was certainly disturbed not only by the exploding shells but also, more important, by the fires they ignited. The fires prevented the growth of woody vegetation and instead maintained sites covered with herbaceous vegetation. Additionally, beavers were firmly established in the impact zone, which included both active and inactive colonies where the former ponds had become grassy wetlands. Such a patchwork produced a fragmented butterfly population at Fort Bragg, which further added to their precarious situation. Indeed, these pockets, together supporting just fifteen hundred of the rare butterflies, slowly blinked out one by one, and by 2010 extinction again loomed. Simply listing the species (in 1994) as endangered clearly was not enough to save it.

At this point, conservation biologists stepped in and began modifying habitat elsewhere at Fort Bragg to create conditions similar to those in the impact zone. The existing regime of prescribed burns was supplemented by the installation of artificial beaver dams (large plastic tubes filled with water), which created new sites of grassy wetlands. In response, some of the rare butterflies immigrated to the new habitat using streams as corridors. In other cases, the new habitat was stocked with butterflies raised in captivity. (These locations remain confidential to lessen the poaching activities of unscrupulous collectors.) The St. Francis' satyr still remains on the federal list of endangered species, but thanks to creative management, its future now seems far more secure than when bombs and beavers alone provided only a small chance for its survival.

READINGS AND REFERENCES

A Vast Plain of Longleaf Pine

Christensen, H. L. 1977. Fire and soil-plant relations in a pine-wiregrass savannah on the Coastal Plain of North Carolina. Oecologia 31:27–44.

Clewell, A. F. 1989. Natural history of wiregrass (*Aristida stricta* Michx., Graminae). Natural Areas Journal 9:223–233.

Earley, L. S. 2004. Looking for Longleaf: The Fall and Rise of an American Forest. University of North Carolina Press, Chapel Hill.

Fowler, J. A. 2015. Orchids, Carnivorous Plants, and Other Wildflowers of the Green Swamp, North Carolina: Exploring America's Most Diverse Ecosystem. Jim Fowler Photography, Greenville, S.C.

Hardin, E. D., and D. L. White. 1989. Rare vascular plant taxa associated with wiregrass (*Aristida stricta*) in the southeastern United States. Natural Areas Journal 9:234–245.

Hare, R. C. 1965. Contribution of bark to fire resistance in southern trees. Journal of Forestry 63:248–251.

Jose, S., E. J. Jokela, and D. L. Miller, eds. 2006. The Longleaf Pine Ecosystem: Ecology, Silviculture, and Restoration. Springer, New York.

Komarek, E. V. 1968. Lightning and lightning fires as ecological forces. Proceedings of the Tall Timber Fire Ecology Conference 8:169–197.

Lipscomb, D. J. 1989. Impacts of feral hogs on longleaf pine regeneration. Southern Journal of Applied Forestry 13:177–181.

Noss, R. F. 1988. The longleaf pine landscape of the Southeast: almost gone and almost forgotten. Endangered Species Update 5(5):1–8.

Walker, J. L. 1993. Rare vascular plant taxa associated with the longleaf pine ecosystems: patterns in taxonomy and ecology. Proceedings of the Tall Timbers Fire Ecology Conference 18:105–126.

Walker, J., and R. K. Peet. 1983. Composition and species diversity of pine-wiregrass savannas of the Green Swamp, North Carolina. Vegetation 55:163–179.

Longleaf Pine Specialists: Red-Cockaded Woodpeckers

Allen, D. H. 1991. An insert technique for constructing artificial red-cockaded woodpecker cavities. General Technique Report SE-73, USDA Forest Service, Ashville, N.C.

Bean, M. J. 2017. Endangered species Safe Harbor Agreements: an assessment working paper. Sand County Foundation / Environmental Policy Innovation Center, Madison, Wis.

Conner, R. N., D. C. Rudolph, D. Saenz, and R. H. Johnson. 2004. The red-cockaded woodpecker cavity tree: a very special pine. Pages 407–411 *in* Proceedings of the Fourth Red-Cockaded Woodpecker Symposium (R. Costa and S. J. Daniels, eds.). Hancock House Publishers, Blaine, Wash.

Edwards, J. W., E. E. Stevens, and C. A. Dachelet. 1997. Insert modifications improve access to artificial red-cockaded woodpecker nest cavities. Journal of Field Ornithology 68:228–234.

Hooper, D. C., and R. N. Conner. 1991. Cavity tree selection by red-cockaded woodpeckers in relation to tree age. Wilson Bulletin 103:458–467.

Hooper, D. C., M. R. Lennartz, and H. D. Muse. 1991. Heart rot and cavity tree selection by red-cockaded woodpeckers. Journal of Wildlife Management 55:323–327.

Jackson, J. A. 1994. Red-cockaded woodpecker: *Picoides borealis*. The Birds of North America, No. 85 (A. Poole and F. Gill, eds.). American Ornithologists' Union, Washington, D.C., and Academy of Natural Sciences, Philadelphia, Pa.

McFarlane, R. W. 1992. A Stillness in the Pines: The Ecology of the Red-Cockaded Woodpecker. Norton, New York.

Rudolph, D. C., H. Kyle, and R. N. Conner. 1990. Red-cockaded woodpeckers vs rat snakes: the effectiveness of the resin barrier. Wilson Bulletin 102:14–22.

Walters, J. R., P. D. Doerr, and J. H. Carter III. 1988. The cooperative breeding system of the red-cockaded woodpecker. Ethology 78:275–305.

Big Cones, Big Squirrels

Kiltie, R. A. 1989. Wildfire and the evolution of dorsal melanism in fox squirrels, *Sciurus niger*. Journal of Mammalogy 70:726–739.

Kiltie, R. A. 1992. Tests of hypotheses on predation as a factor maintaining polymorphic melanism in coastal-plain fox squirrels (*Sciurus niger* L.). Biological Journal of the Linnean Society 45:17–37.

Long, K. 1995. Squirrels: A Wildlife Handbook. Johnson Books, Boulder, Colo. [Includes Bachman's quote in a letter to Audubon.]

Weigl, P., M. A. Steele, L. J. Sherman, et al. 1989. The ecology of the fox squirrel (*Sciurus niger*) in North Carolina: implications for survival in the Southeast. Bulletin 24, Tall Timbers Research Station, Tallahassee, Fla.

Venomous and Vulnerable

Adkins Giese, C. L., D. N. Greenwald, D. B. Means, et al. 2011. Petition to list the eastern diamondback rattlesnake (*Crotalus adamanteus*) as threatened under the endangered species act. Center for Biological Diversity, Portland, Ore. [Thorough review of the biology, habitat, threats, and status of the species.]

Martin, W. H., and D. B. Means. 2000. Geographical distribution and habitat relationships of the eastern diamondback rattlesnake, *Crotalus adamanteus*. Herpetological Natural History 7:9–34.

Means, D. B. 2005. The value of dead tree bases and stump holes as habitat for wildlife. Pages 74–78 *in* Amphibians and Reptiles: Status and Conservation in Florida (W. E. Meshaka Jr. and K. J. Babbitt, eds.). Krieger Publications, Malabar, Fla.

Means, D. B. 2009. Effects of rattlesnake roundups on the eastern diamondback rattlesnake (*Crotalus adamanteus*). Herpetological Conservation and Biology 4:132–141.

Carnivorous Plants

Adlassnig, W., M. Peroutka, and T. Lendl. 2011. Traps of carnivorous pitcher plants as a habitat: composition of the fluid, biodiversity, and mutualistic activities. Annals of Botany 107:181–194.

Dahlem, G. A., and R. F. C. Naczi. 2006. Flesh flies (Diptera: Sarcophagidae) associated with North American pitcher plants (Sarraceniaceae), with descriptions of three new species. Annals of the Entomological Society of America 99:218–240.

Darwin, C. 1875. Carnivorous Plants. John Murray, London. [2010 reprint, New York University Press, New York.]

Davis, A. L., M. H. Babb, M. C. Lowe, et al. 2019. Testing Darwin's hypothesis about the wonderful Venus flytrap: marginal spikes form a "horrid prison" for moderate-sized insect prey. American Naturalist 193:309–317.

Forterre, Y., J. M. Skotheim, J. Dumais, and L. Mahadevan. 2005. How the Venus flytrap snaps. Nature 433:421–425.

Gibson, T. C., and D. M. Waller. 2009. Evolving Darwin's "most wonderful plant": ecological steps to a snap trap. Phytologist 183:575–587.

Jennings, D. E., J. J. Krupa, T. R. Raffel, and J. R. Rohr Jr. 2010. Evidence for competition between carnivorous plants and spiders. Proceedings of the Royal Society B, Biological Sciences 277:3001–3005.

Stephens, J. D., and D. R. Folkerts. 2012. Life history aspects of *Exyra semicrocea* (pitcher plant moth) (Lepidoptera: Noctuidae). Southeastern Naturalist 11:111–126.

Turner, M. 1979. Diet and feeding phenology of the green lynx spider, *Peucetia viridans* (Araneae: Oxyopidae). Journal of Arachnology 7:149–154.

Inland Dunes: The Carolina Sandhills

Anonymous. 2018. Conservation plan for the gopher frog (*Rana capito*) in North Carolina. North Carolina Wildlife Resources Commission, Raleigh.

Braham, R., C. Murray, and M. Boyer. 2006. Mitigating impacts to Michaux's sumac (*Rhus michauxii*): a case study of transplanting an endangered shrub. Castanea 71:265–271.

Burger, J., and R. T. Zappalorti. 1986. Nest site selection by pine snakes, *Pituophis melanoleucus*, in the New Jersey Pine Barrens. Copeia 1986:116–121.

Burger, J., and R. T. Zappalorti. 1991. Nesting behavior of pine snakes (*Pituophis m. melanoleucus*) in the New Jersey Pine Barrens. Journal of Herpetology 25:152–160.

Burger, J., and R. T. Zappalorti. 2011. The northern pine snake (*Pituophis m. melanoleucus*) in New Jersey: its life history, behavior, and conservation. Pages 1–56 *in* Reptiles: Biology, Behavior and Conservation (K. J. Baker, ed.). Nova Science Publishers, Hauppauge, N.Y.

Burger, J., R. T. Zappalorti, J. Dowdell, et al. 1992. Subterranean predation on pine snakes (*Pituophis melanoleucus*). Journal of Herpetology 26:259–263.

Freda, J., and R. J. Gonzalez. 1986. Daily movements of the treefrog, *Hyla andersonii*. Journal of Herpetology 20:469–471.

Hall, J. 2020. A frog on the brink. Wildlife in North Carolina 84(1): 30–37. [Describes efforts to improve the perilous status of gopher frogs.]

Hansen, K. L. 1958. Breeding pattern of the eastern spadefoot toad. Herpetologica 14:57–67.

Hulmes, D., P. Hulmes, and R. Zappalorti. 1981. Notes on the ecology and distribution of the Pine Barrens treefrog, *Hyla andersonii*, in New Jersey. Bulletin of the New York Herpetological Society 17:2–19.

Humphries, W. J., and M. A. Sisson. 2012. Long distance migrations, landscape use, and vulnerability to prescribed fire of the gopher frog (*Lithobates capito*). Journal of Herpetology 46:665–670.

Krabbenhoft, T. J., R. C. Rohde, A. N. Leibman, and J. M. Quattro. 2008. Concordant mitochondrial and nuclear DNA partitions define Evolutionarily Significant Units in the imperiled pinewoods darter, *Etheostoma mariae* (Pices: Percidae). Copeia 2008:909–915.

Means, D. B., and C. J. Longden. 1976. Aspects of the biology and zoogeography of the Pine Barrens treefrog (*Hyla andersonii*) in northern Florida. Herpetologica 32:117–130.

Morin, P. J., S. P. Lawler, and E. A. Johnson. 1990. Ecology and breeding phenology of larval *Hyla andersonii*: the disadvantages of breeding late. Ecology 71:1590–1598.

Palis, J. G. 1998. Breeding biology of the gopher frog, *Rana capito*, in western Florida. Journal of Herpetology 32:217–223.

Pearson, P. G. 1955. Population ecology of the spadefoot toad, *Scaphiopus h. holbrooki* (Harlan). Ecological Monographs 25:234–267.

Pehek, E. L. 1995. Competition, pH, and the ecology of larval *Hyla andersonii*. Ecology 76:1786–1793. [Experiments designed to determine why Pine Barrens treefrogs only occur in highly acidic wetlands.]

Rambert, D. H., Jr. 1979. The Carolina plants of Andre Michaux. Castanea 44:65–80.

Rohde, F. C., and R. G. Arndt. 1997. Rare fishes of North Carolina (part one). American Currents 23:4–8. [Concerns species endemic to the Carolina Sandhills.]

Rohde, F. C., and S. W. Ross. 1987. Life history of the pinewoods darter,

Etheostoma mariae (Osteichthyes: Percidae), a fish endemic to the Carolina Sandhills. Brimleyana 3:1–20.

Semlitsch, R. D., J. W. Gibbons, and T. D. Tuberville. 1995. Timing of reproduction and metamorphosis in the Carolina gopher frog (*Rana capito capito*) in South Carolina. Journal of Herpetology 29:612–614.

Sorrie, B. A. 2011. A Field Guide to the Wildflowers of the Sandhills Region. University of North Carolina Press, Chapel Hill.

Swezey, C. S., B. A. Fitzwater, G. R. Whittecar, et al. 2016. The Carolina Sandhills: Quaternary Eolian sand sheets and dunes along the updip margin of the Atlantic Coastal Plain province, southeastern United States. Quaternary Research 86:271–286.

Wells, B. W. 1929. A new pixie from North Carolina. Journal of the Elisha Mitchell Scientific Society 44:238–239.

Wells, B. W., and I. V. D. Shunk. 1931. Vegetation and habitat factors of the coarser sands of the North Carolina Coastal Plain. Ecological Monographs 1:465–520. [Describes the experiment with oak leaves.]

Woolcott, W. S., and E. G. Maurakis. 1988. Pit-ridge nest construction and spawning behaviors of *Semotilus lumbee* and *Semotilus thoreauianus*. Proceedings Southeastern Fishes Council 18:1–3.

Bombs, Beavers, and Butterflies

Clayton, H., N. M. Haddad, B. Ball, et al. 2015. Habitat restoration as a recovery tool for a disturbance-dependent butterfly, the endangered St. Francis' satyr. Pages 147–159 *in* Butterfly Conservation in North America (J. C. Daniels, ed.). Springer, New York.

Haddad, N. M. 2018. Resurrection and resilience of the rarest butterflies. PLOS Biology 16(2):e2003488. https://doi.org/10.1371/journal.pbio.2003488. [A memoir featuring a "eureka moment" about the importance of disturbed habitat for St. Francis' satyrs and other rare butterflies.]

Kuefler, D., N. M. Haddad, S. Hall, et al. 2008. Distribution, population structure, and habitat use of the endangered St. Francis' satyr butterfly. American Midland Naturalist 159:298–320.

Milko, L. V., N. M. Haddad, and S. L. Lance. 2012. Dispersal via stream corridors structures populations of the endangered St. Francis' satyr butterfly (*Neonympha mitchellii francisci*). Journal of Insect Conservation 16:263–273.

Parshall, D. K., and T. W. Kral. 1989. A new subspecies of *Neonympha mitch-ellii* (French) (Satyridae) from North Carolina. Journal of the Lepidopterists' Society 43:114–119.

Infobox 6-1: Naval Stores: Wooden Ships and Tar Heels

Bennett, A. B. 2000. Assessment of the impact of raking pine straw on arthropods in longleaf pine (*Pinus palustris*) forests. MS thesis, University of North Carolina at Wilmington.

Hodge, A. W. 2006. The naval stores industry. Pages 43–48 *in* The Longleaf Pine Ecosystem: Ecology, Silviculture, and Restoration (S. Jose, E. J. Jokela, and D. L. Miller, eds.). Springer, New York.

Outland, R. B., III. 2004. Tapping the Pines: The Naval Stores Industry in the American South. Louisiana State University Press, Baton Rouge.

Watson, A. D. 1992. Wilmington: Port of North Carolina. University of South Carolina Press, Columbia. [Describes the production and shipping of naval stores.]

Infobox 6-2: Fire Ecology in Brief

Cooper, C. F. 1961. The ecology of fire. Scientific American 204:150–160.

Guyer, C., and M. A. Bailey. 1993. Amphibians and reptiles of longleaf pine communities. Proceedings of the Tall Timbers Fire Ecology Conference 18:139–158.

Higgins, K. F. 1984. Lightning fires in North Dakota grasslands and in pine-savannah lands of South Dakota and Montana. Journal of Range Management 37:100–103.

Innes, R. J. 2009. *Gopherus polyphemus*. Unnumbered pages *in* Fire Effects Information System, www.fs.fed.us/database/feis/animals/reptile/gopo /all.html. USDA Forest Service, Rocky Mountain Experiment Station, Fort Collins, Colo.

Lee, J., J. Brumley, M. Ryckeley, et al. 2016. Pitcher plant moths (*Exyra*) fly from pitchers in response to smoke. Journal of the Lepidopterists' Society 70:268–270.

Lynch, J. J. 1941. The place of burning in the management of the Gulf Coast refuges. Journal of Wildlife Management 5:454–457.

Miller, K. V., B. R. Chapman, and K. K. Ellington. 2001. Amphibians in pine stands managed with growing-season and dormant-season prescribed fires. Journal of the Elisha Mitchell Scientific Society 117:75–78.

Singer, F. J., W. Schreier, J. Oppenheim, and E. O. Garton. 1989. Drought, fires, and large mammals: estimating the 1988 severe drought and large-scale fires. BioScience 39:716–722. [Describes animals killed by the fires.]

Wright, H. A., and A. W. Bailey. 1982. Fire Ecology: United States and Southern Canada. John Wiley & Sons, New York.

Infobox 6-3: Old Logs, New Floors

Davis, W. 2001. Old growth gold. StarNews (Wilmington, N.C.), April 11, 2001, 1A, 5A.

Earley, L. S. 2004. Looking for Longleaf: The Fall and Rise of an American Forest. University of North Carolina Press, Chapel Hill. [See page 151 for estimates of lost "sinkers."]

Elmore, C. J. 2004. Sunken treasure. Business, North Carolina 24(11):28.

Fowler, M. 1940. The last of the whitewater timber rafters. The State 8(1):6–7. [Precursor to Our State.]

Greene, A. 1999. Loggers take plunge for best lumber. StarNews (Wilmington, N.C.), July 11, 1999, 1E, 4E.

Trout, W. E., III, J. Hairr, and N. R. Trout. 2016. The Cape Fear River Atlas. Virginia Canals and Navigations Society, Madison Heights, Va. [See pages 86–87 for more about rafting logs.]

To keep every cog and wheel is the first precaution of intelligent tinkering.—ALDO LEOPOLD, *Round River: From the Journals of Aldo Leopold*

Afterword

North Carolina unfortunately is well represented when it comes to endangered species, for which just a few statistics reflect the grim circumstances.[1] The federal list for North Carolina tallies seventeen species of endangered plants and twenty-four species of endangered animals. Of these, 41 percent and 42 percent, respectively, occur in the Coastal Plain.[2] Strategies to improve the status of each vary considerably, among them habitat improvements, supplementing wild populations with stock from captive breeders, and dedicated sanctuaries, and while some management details are mentioned in the previous pages, further discussions of these deserve fuller attention elsewhere. However, three overarching questions about endangered species fall within the scope of this book. First, what factors underlie their dire circumstances? Second, why are some species so vulnerable while others are not? Finally, why should we care?

[1] The Endangered Species Act of 1973 recognizes species, subspecies, and imperiled populations of otherwise abundant taxa as "species" for purposes of its mandate. Here we follow the same usage of "species."
[2] State-wide, threatened species include nine plants and eleven animals, of which 22 percent and 91 percent, respectively, occur on the Coastal Plain.

What Factors Cause Extinctions?

1. NATURAL CAUSES. Throughout time, species have evolved only to later disappear, often as the result of catastrophic events such as volcanic eruptions, abrupt climatic changes, or, notably, the impact of a huge asteroid. Others simply disappeared because they evolved into newer forms or became so overspecialized that they could not compete with more generalized forms. Diseases no doubt also played a role in some cases. Based on the fossil record, a bird endures as a species for an average of about two million years, whereas a mammal exists for just six hundred thousand years. Until AD 1600, when profound human influences came to the fore, natural causes claimed an average of about five species per year. Natural disasters continue to affect species, but they hardly account for the current wave of extinctions, which globally may number one per hour—or even more.

2. WANTON EXPLOITATION. Unregulated hunting brought an end to passenger pigeons, once the most abundant bird in North America, in large part because the slaughter focused on their nesting colonies. The same fate befell Carolina parakeets and nearly claimed bison and, supplemented by traps and poisons, gray wolves. Some species of whales were also brought to the brink by relentless exploitation.

3. INTRODUCED PREDATORS AND COMPETITORS. Introduced organisms, often identified as "exotic" species, may invade and alter the structure and function of natural communities, often to the detriment of native species. Monotypic stands of common cane, for example, virtually eliminate the natural biodiversity in large areas of salt marsh, and European starlings usurp the cavities where eastern bluebirds nest. Red imported fire ants eradicate local populations of native ants and seriously impair the reproductive success of ground-nesting birds. Burmese pythons continue to reduce mammal populations in the Everglades.

4. HABITAT MODIFICATION. Loss of suitable habitat tops the list of threats to plants and animals. Humans, of course, serve as the primary agents for destroying soil, water, and vegetation. Drainage projects, reservoir construction, deforestation, pollution, and industrial,

agricultural, and residential development are just some of the more obvious human-induced changes that become the ultimate destroyers of species. In North Carolina, the steady destruction of the once vast longleaf pine forest perhaps offers the best example of lost habitat, closely followed by the loss of wetlands and maritime forests.

Why Are Some Species So Vulnerable?

Some species—coyotes, for example—defy long-standing efforts to reduce their numbers, yet others are far less able to adapt to adversities that may lead to extinction. Stated otherwise, what features predispose certain species to the perils of extinction? Two or more features may be involved in some instances, but just one may be more than enough.

1. Species with narrow habitat requirements or restricted distributions. Red-cockaded woodpeckers represent the former, and some fishes and mussels at Lake Waccamaw are examples of the latter. Species limited to oceanic islands are particularly vulnerable.

2. Species of importance for their flesh, fur, feathers, or ivory, among them spotted cats and egrets, and others for body parts with alleged medicinal benefits, notably rhinoceros horns. As a particularly egregious example from the eighteenth century, extinction claimed Steller's sea cows just twenty-seven years after their discovery, victims of hunters seeking the hides, buttery fat, and meat so amply provided by these eight-to-ten-ton pinnipeds. Moreover, species spending all or part of their lives in international waters lack legal protection, as is the case for whales and sea turtles.

3. Species of large size, especially predators or those intolerant of humans (or vice versa), as well as those requiring large ranges, for which grizzly bears and gray wolves serve as prime examples. Mountain lions vanished in North Carolina as human populations expanded across the state.

4. Species with limited reproductive potential resulting from one or more of the following: small numbers of offspring per breeding

season, long gestation or incubation periods, extensive parental care, and slow to reach breeding age. Giant pandas, the international icon for endangered species, share several of these features, as do whales.

5. Species with genetic vulnerability, for which red wolves coming in contact with coyotes offer a classic example, as discussed more fully in Chapter 5. Mallards similarly "swamp" the gene pool of American black ducks.

6. Species with specialized physical, behavioral, or physiological features, the latter shown by the cold sensitivity of Florida manatees. In North Carolina, individuals lingering too long in autumn suffer cold stress and may die when water temperatures drop below sixty-eight degrees. Atlantic right whales, because their migration route lies just offshore, become fatally entangled with fishing gear or collide with ships.

Why Care about Endangered Species?

Nearly everyone has a good measure of empathy concerning the welfare of whales, whooping cranes, giant pandas, and stately redwoods. These, along with a few others, typify charismatic species that bring forth a feeling akin to "cuddly," "majestic," or "awesome." In short, such species are an "easy sell" for protection and proactive management. But similar efforts for many other endangered species, perhaps a spider, a snake, a minnow, or an obscure fern, often garner no more than a shrug and a dismissive "Who cares?" The thoughts and lives of three titans of conservation, each with different approaches to a common cause, offer a response.

Scottish-born John Muir (1838–1914), after studying geology and chemistry in Wisconsin, decided to see the wonders of North America and set off on foot for the Gulf of Mexico on what turned out to be a one-thousand-mile trek. From there, he sailed to California, where the grandeur of the Sierra Nevada, particularly Yosemite Valley, quickly captivated the doughty Scot. Conversely, wholesale logging of the huge redwoods dismayed Muir, who responded by writing newspaper articles and, later, books about his beloved mountains and the threats befalling their forest ecosystems. His relentless campaign for preserving wilderness became the mission of the Sierra Club, which he founded in 1892, and later captured the interest of a young president named Theodore Roosevelt. Indeed, Roosevelt traveled to California, where he and Muir camped alone for three days in the redwood wilderness. Awed by the experience, Roosevelt declared the giant trees as "monuments in themselves" worthy of preservation—and did just that when he created Muir Woods National Monument in 1908. In 1911, Muir proposed formation of an agency to oversee national parks and monuments, but then President William Howard Taft failed to act on the idea. However, in 1916, two years after Muir's death, his vision became a reality with the creation of the National Park Service.

Muir held a deep-seated belief that nature is a temple often degraded by the economic activities of humans—a view rooted in the concepts of Ralph Waldo Emerson and Henry David Thoreau, both prominent philosophers of the mid-1800s. Their thoughts emerged as the Romantic-Transcendental Preservation Ethic, a hefty name for the idea that a communion with nature brought humans closer to their divine creator. Harmony and the beauty of nature form the bedrock of transcendentalism, and for Muir this meant preserving wilderness whole and intact. In his view, species must be preserved because they represent the marvelous works of God, and their continued existence must not be threatened by the work of thoughtless and often greedy humans. Were Muir alive today, he would champion the Endangered Species Act as a blessing that protects God's creations.

Next we turn to forester Gifford Pinchot (1865–1946). Born into wealth, Pinchot studied botany, geology, and meteorology at Yale, then studied forestry in Europe. (Forestry had not yet become an academic field in the United States.) In doing so, he emerged as the first American trained as a professional forester. Pinchot came to the attention of George W. Vanderbilt, who was building an estate, Biltmore, on a large tract of neglected forest in the mountains of North Carolina. (The land, in part, later became the nucleus for Pisgah National Forest.) Vanderbilt hired Pinchot in 1892 not only to improve and manage the forest for timber production but also to prepare an exhibition and literature about the

Biltmore Forest for the World's Columbian Exposition in 1893. In all, Pinchot's efforts produced favorable publicity about scientific forestry in America. He stayed at Biltmore until 1895; his successor, German-born forester Carl A. Schenck, established the Biltmore Forest School, a rigorous vocational program commonly regarded as the birthplace of American forestry. (In 1900, Pinchot helped found the Yale School of Forestry, the first academic program of its kind in the United States.)

Pinchot worked as a consultant for a year, then accepted a position with the government to survey the nation's forest reserves. His professionalism did not go unnoticed, and when President Roosevelt created the U.S. Forest Service in 1905, he appointed Pinchot as its first chief. The two men worked closely together to withdraw large areas from the public domain to create national forests. Along the way, Pinchot coined the term "conservation," which quickly entered the lexicon of Roosevelt and others concerned with the welfare of natural resources. Pinchot's influence waned after Roosevelt left office, and he resigned during the administration of President Taft, who did not share their progressive views. Although twice running unsuccessfully for the Senate, Pinchot won two terms as governor of Pennsylvania and later served as an advisor to Presidents Franklin Roosevelt and Harry Truman.

Pinchot established the Resource Conservation Ethic, which emphasizes the economic values of forests and other natural resources and thereby differed greatly from Muir's spiritual concepts of nature. In Pinchot's view, forests, grasslands, and other natural areas warrant good stewardship because of the services and products they yield for human use. This narrows the importance of many species and may completely overlook many others because it fails to recognize not only intangible values but also the potential discovery of products from species currently regarded as "worthless." As one example, the bark of Pacific yew provided the initial source of a drug that successfully treated certain types of cancer. Yet because its timber value is limited, Pacific yew was for decades regarded as a weed to be removed in favor of other species better suited for timber or paper production. Had the species vanished, so too would have the future discovery of its medicinal properties, which underscores the moral that the only truly worthless species is one that has become extinct. Nonetheless, Pinchot deserves lasting credit for insisting that forests be managed on a sustained basis, and in doing so, he prevented their wanton destruction for quick profits and a ruinous future. Stated otherwise, we can ascribe to him the idea of renewability for forests and, by extension, many other natural resources, among them soil, water, and wildlife. Moreover, with Roosevelt's backing, Pinchot placed huge areas under federal protection, many of which provide critical habitat for species threatened with extinction.

Final mention goes to Aldo Leopold (1887–1948), the father of wildlife management in the United States and author of *A Sand County Almanac*, a collection of essays still revered in the pantheon of conservation literature. Leopold graduated from the Yale School of Forestry in 1909 and thereafter conducted timber surveys in Arizona and New Mexico for the fledgling U.S. Forest Service. These experiences provided him with an appreciation of untrammeled landscapes, which he believed should be protected from any form of exploitation. Because of Leopold's persistent lobbying, the first national wilderness areas within a national forest became a reality—no mean accomplishment, given the continuing predominance of Pinchot's views about timber production in national forests. In fact, decades would pass before the National Wilderness Preservation Act became public policy in 1964. By 1933, he had left the U.S. Forest Service and joined the faculty at the University of Wisconsin, where he remained until his death; he published *Game Management*, the first textbook of its kind, in the same year. Under Leopold's influence, wildlife management came of age just as forestry had matured under Pinchot. In 1935, Leopold cofounded The Wilderness Society and, in 1937, The Wildlife Society.

In 1935, Leopold acquired a tract of exhausted farmland, which included a rustic shack, where he retreated to write his essays and worked to restore the land.[3] In *A Sand County Almanac* (so named for the

3   Leopold died fighting a grassfire near his beloved shack in 1948, a year before *Almanac* appeared. The book has remained in continuous publication ever since. However, the quote about "intelligent tinkering" appears elsewhere; see *Round River: From the Journals of Aldo Leopold*, published in 1953 by one of his sons, Luna B. Leopold.

sandy soils of Sauk County, Wisconsin, where the property was located), Leopold reflected on his close touch with the forested wilderness in the American Southwest and the impoverished land at his hideaway in Wisconsin. These experiences culminated in the Evolutionary-Ecological Land Ethic, which views nature as a complex of interlocking parts that have evolved into a functioning, interdependent system, or what is often referred to as the web of life. In his view, each species has its own intrinsic value aside from any tangible or economic value it may have for humans. Species thus lack positive, neutral, or negative values and instead attain their worth as parts of an ecosystem in which humans are but one part and not its masters. Humans might use and manage nature, but only in ways that pose no harm to the health of ecosystems. For example, in the essay "Thinking Like a Mountain," Leopold famously revealed his own epiphany about the worth of predators. He initially regarded wolves as a scourge for deer and hence thought they should be shot, trapped, or poisoned at every opportunity, but he later realized that predators were keystones that kept herbivores from overeating their food supply and, ultimately, reducing biological diversity. The concept later became formally recognized by ecologists as a top-down trophic cascade and led to the reintroduction of wolves into Yellowstone National Park and the recovery of an ecosystem degraded by too many elk.

Someone once observed that losing a species is akin to burning a book before it has been read, to which we might add that losing an ecosystem—longleaf pine forests, for example—is like burning an entire library. Whether we favor the thoughts of Muir, Pinchot, or Leopold makes little difference, but it seems essential to recognize the overarching message each conveys. As summarized by the late John Sawhill, former president and CEO of The Nature Conservancy, "A society is defined not only by what it creates, but also by what it refuses to destroy." Somewhere, perhaps under a decaying log or in a shimmering pond, the fate of an obscure creature rests precisely on that premise.

Bolen, E. G. 2006. Why endangered species matter. Wildlife in North Carolina 70(11):26–30. [Elements of this essay were adapted for the current text.]

Eisenberg, C. 2010. The Wolf's Tooth: Keystone Predators, Trophic Cascades, and Biodiversity. Island Press, Washington, D.C.

Fisher, J., N. Simon, and J. Vincent. 1969. Wildlife in Danger. Viking Press, New York. [A primary source for this chapter. A large volume of literature has since appeared about extinction rates, often with different assessments. All agree, however, that human influences have greatly accelerated the extinction of Earth's biological resources. In contrast with those for animals, extinction rates for plants remain elusive.]

Flader, S. L. 1974. Thinking Like a Mountain: Aldo Leopold and the Evolution of an Ecological Attitude toward Deer, Wolves, and Forests. University of Nebraska Press, Lincoln.

Leopold, A. 1933a. The conservation ethic. Journal of Forestry 31:634–643.

Leopold, A. 1933b. Game Management. Charles Scribner's Sons, New York.

Leopold, A. 1949. A Sand County Almanac. Oxford University Press, New York. [A "must read" for conservationists.]

Leopold, L. B., ed. 1953. Round River: From the Journals of Aldo Leopold. Oxford University Press, New York.

Meine, C. 1988. Aldo Leopold, His Life and Work. University of Wisconsin Press, Madison.

Pinkett, H. T. 1957. Gifford Pinchot at Biltmore. North Carolina Historical Review 34:346–357.

Pinkett, H. T. 1970. Gifford Pinchot, Private and Public Forester. University of Illinois Press, Urbana.

Ripple, W. J., R. L. Beschta, and L. E. Painter. 2015. Trophic cascades from wolves to alders in Yellowstone. Forest Ecology and Management 354:254–260.

Schenck, C. A. 2011. Cradle of Forestry in America: Biltmore Forest School, 1898–1913. Forest History Society, Durham, N.C.

Wani, M. C., H. L. Taylor, M. E. Wall, et al. 1971. Plant antitumor agents IV. The isolation and structure of taxol, a novel antileukemic and antitumor agent from Taxus brevifolia. Journal of the American Chemical Society 93:2325–2327.

Worster, D. 2008. A Passion for Nature: The Life of John Muir. Oxford University Press, New York.

Appendix.
Latin Names of Organisms

This list presents the scientific names for organisms mentioned in the text. In some cases, research based on molecular genetics has led to changes in long-standing names; hence, some may not seem familiar or agree with those appearing in literature citations. To clarify these situations, the vacated name appears in brackets immediately following the currently accepted name. For example, smooth cordgrass appears as *Sporobolus alterniflorus* [*Spartina alterniflora*]. Brief parenthetical annotations accompany entries where further information may be helpful. An asterisk (*) designates an extinct species.

Plants, including Mosses, Algae, and Fungi

Atlantic white cedar, *Chamaecyparis thyoides*

Bald cypress, *Taxodium distichum*

Bayberry, *Morella* [*Myrica*] *pensylvanica*

Beach bean, *Strophostyles helvola* (the "pea" of Pea Island)

Beach vitex, *Vitex rotundifolia* (introduced invasive)

Beechnut (American beech), *Fagus grandifolia*

Black needlerush, *Juncus roemerianus*

Black walnut, *Juglans nigra*

Bladderwort, *Utricularia* spp.

Blueberry, *Vaccinium* spp.

Blueflower butterwort, *Pinguicula caerulea*

Brown-spot needle blight (fungus), *Mycosphaerella dearnessii*

Bull nettle, *Cnidoscolus stimulosus*

Bulrush, *Scirpus* spp. (revision of the bulrush group now includes genera *Schoenoplectus* and *Bolboschoenus*, which apply to some species formerly in the genus *Scirpus*)

Bur-reed, *Sparganium americanum*

Bushy pondweed (southern naiad), *Najas guadalupensis*

Buttonbush, *Cephalanthus occidentalis*

Cape Fear spatterdock, *Nuphar sagittifolia*

Carolina ash, *Fraxinus caroliniana*

Carolina dropseed, *Sporobolus pinetorum*

Carolina lily, *Lilium michauxii*

Carolina sandwort, *Minuartia caroliniana*

Cattail, *Typha* spp.

Cheatgrass, *Bromus tectorum* (introduced invasive)

Cherrybark oak, *Quercus pagoda*

Coastal bluestem, *Schizachyrium littorale*

Coastal morning glory, *Ipomoea cordatotriloba*

Cocklebur, *Xanthium strumarium*

Colic root, *Aletris farinosa*

Common reed, *Phragmites australis* [*communis*]

Dandelion, *Taraxacum* spp.

Dixie reindeer lichen, *Cladonia subtenuis* (one of the more common species in the Sand Barrens; this and related species are often and incorrectly called a "moss")

Duckweed, *Lemna* spp.

Dune camphorweed, *Heterotheca subaxillaris*

Dune hairgrass (sweetgrass), *Muhlenbergia sericae*

Dwarf palmetto, *Sabal minor*

Dwarf sundew, *Drosera brevifolia*

Eelgrass, *Zostera marina*

Elms, *Ulmus* spp.

Eurasian watermilfoil, *Myriophyllum spicatum*

Fetterbush, *Lyonia lucida*

Franklin's lost tree, *Franklinia alatamaha*

Gallberry holly, *Ilex glabra*

Giant cane, *Arundinaria gigantea*

Giant cordgrass, *Sporobolus* [*Spartina*] *cynosuroides*

Giant cutgrass, *Zizaniopsis miliacea*

Green ash, *Fraxinus pennsylvanica*

Hackberry (sugarberry), *Celtis* spp.

Hickory, *Carya* spp.

Honey cup, *Zenobia pulverulenta*

Huckleberry, *Gaylussacia frondosa*

Hydrilla, *Hydrilla verticillata* (introduced invasive)

Ironwood, *Carpinus caroliniana*

Jack pine, *Pinus banksiana*

Laurel greenbriar, *Smilax laurifolia*

Laurel oak, *Quercus laurifolia*

Least trillium, *Trillium pusillum*

LeConte's thistle, *Cirsium lecontei*

Littleleaf pixie-moss, *Pyxidanthere brevifolia*

Loblolly bay, *Gordonia lasianthus*

Loblolly pine, *Pinus taeda*

Maple, *Acer* spp.

Maritime bushy bluestem, *Andropogon tenuispatheus*

Marsh hay (saltmarsh hay; saltmeadow cordgrass),
Sporobolus pumilus [*Spartina patens*]

Marsh pink, *Sabatia stellaris*

Michaux's sumac, *Rhus michauxii*

Milkweed, *Asclepias* spp.

Mitchell's sedge, *Carex mitchelliana*

Mulberry, *Morus* spp.

Muskgrass, *Chara* spp.

Nutmeg hickory, *Carya myristiciformis*

Oak, *Quercus* spp.

Orange milkwort (red-hot poker), *Polygala* [*Pylostachya*] *lutea*

Pawpaw, *Asimina triloba*

Pennywort, *Hydrocotyle bonariensis*

Perennial saltmarsh aster, *Symphyotriichum* [*Aster*] *tenuifolium*

Pickerelweed, *Pontederia cordata*

Pickleweed, *Salicornia* spp.

Pineland scalypink (wire plant), *Stipulicida setacea*

Pink sundew, *Drosera capillaris*

Pitch pine, *Pinus rigida*

Pond pine, *Pinus serotina*

Purple pitcher plant, *Sarracenia purpurea*

Purple sandgrass, *Triplasis purpurea*

Red heart fungus, *Phellinus pini*

Red maple, *Acer rubrum*

Red tide, *Karenia brevis* (a dinoflagellate alga)

Redwood, *Sequoia sempervirens*

Ribbonleaf pondweed, *Potamogeton epihydrus*

Rosebud orchid, *Cleistesiopsis* [*Pogonia*; *Cleistes*] *divaricata*

Roughleaf dogwood, *Cornus asperifolia*

Rush featherline, *Pleea tenuifolia*

Sago pondweed, *Potamogeton pectinatus*

Sand spikemoss (spiny spikemoss), *Bryodesma* [*Selaginella*] *acanthonota*

Sargasso seaweed, primarily *Sargassum natans* and *S. fluitans* (both brown algae)

Sarvis holly, *Ilex amelanchier*

Savanna iris, *Iris tridentate* [*tripetala*]

Seabeach amaranth, *Amaranthus pumilus*

Sea lavender, *Limonium carolinianum*

Sea oats, *Uniola paniculata*

Sea oxeye, *Borrichia frutescens*

Seaweeds, Rhodophyta, Chlorophyta, and Phaeophyaceae (taxa of macroalgae)

Shagbark hickory, *Carya ovata*

Shoal grass, *Hadodule beaudettei* [*H. wrightii*; *Diplantheria wrightii*]

Silky camellia, *Stewartia malacodendron*

Smooth cordgrass, *Sporobolus alterniflorus* [*Spartina alterniflora*]

Southern maidenhair (Venus hair fern), *Adiantum capillus-veneris*

Southern seaside goldenrod, *Solidago mexicana*

Southern twayblade, *Listera australis* [*Neottia bifolia*]

Spanish moss, *Tillandsia usneoides* (not a moss)

Spoonleaf sundew, *Drosera intermedia*

Stonewort, *Nitella flexilis* (alga)

Sugarberry (hackberry), *Celtis laevigata*

Sundew, *Drosera* spp.

Swamp chestnut oak, *Quercus michauxii*

Sweetgum, *Liquidambar styraciflua*

Sweet pepperbush, *Clethra alnifolia*

Sycamore, *Platanus occidentalis*

Threadleaf sundew, *Drosera filiformis*

Three-square bulrush, *Schoenoplectus [Scirpus] americanus*

Titi, *Cyrilla racemiflora*

Tupelo gum, *Nyssa aquatica*

Turkey oak, *Quercus laevis*

Venus flytrap, *Dionaea muscipula*

Waterwheel plant, *Aldrovanda vesiculosa*

Wax myrtle, *Morella [Myrica] cerifera*

White pine, *Pinus strobus*

White-topped sedge, *Rhynchospora [Dichromena] latifolia*

White wicki, *Kalmia cuneata*

Widgeongrass, *Ruppia maritima*

Wild celery, *Vallisneria americana*

Willow oak, *Quercus phellos*

Wiregrass, *Aristida stricta*

Yaupon, *Ilex vomitoria*

Yellow pitcher plant, *Sarracenia flava*

Yellow waterlily (spatterdock), *Nuphar luteum*

Invertebrates

Atlantic oyster, *Crassostrea virginica*

Atlantic pigtoe, *Fusconaia masoni*

Black flies, family Simuliidae

Blue crab, *Callinectes sapidus*

Caddisflies, order Trichoptera

Cape Fear spike, *Elliptio marsupiobesa*

Chigger (red bugs), *Trombicula alfreddugesi*

Copepods, order Copepoda

Coquina, *Donax variabilis*

Crayfish, families Astacidae and Cambaridae

Crystal skipper, *Atryonopsis quinteri*

Damselflies, order Odonata

Dragonflies, order Odonata

Eastern mud snail, *Tritia [Ilyanassa] obsoleta*

Eastern tiger swallowtail, *Papilio glaucus*

Eelgrass limpet, *Lottia alveus**

Eel swim-bladder nematode, *Anguillicola crassus*

Fingernail clams (pea clams), Sphaeriidae

Flies, order Diptera

Funnel-web wolf spider, *Sosippus floridanus*

Gooseneck barnacle, *Lepas anatifera*

Greenhead. *See* Saltmarsh greenhead

Green lynx spider, *Peucetia viridans*

Honey bee, *Apis mellifera* (introduced)

Horseshoe crab, *Limulus polyphemus*

Jumping spider, *Pelegrina* [*Metaphidippus*] *tillandsiae*
   (a species strongly associated with Spanish moss)

Magnificent ramshorn snail, *Planorbella magnifica*

Marsh periwinkle, *Littoraria* [*Littorina*] *irrorata*

Mayflies, order Ephemeroptera

Midges, family Chironomidae

Monarch butterfly, *Danaus plexippus*

Mud fiddler crab, *Uca pugnax*

Native southern fire ant, *Solenopsis xyloni*

Palamedes butterfly, *Papilio palamedes*

Pea crab, *Pinnotheres ostreum* [*Zaops ostreus*]

Pitcher plant moth, *Exyra* spp. (*E. semicrocea* is a generalist,
   whereas *E. fax* relies exclusively on purple pitcher plants)

Pod lance, *Elliptio folliculata*

Red imported fire ant, *Solenopsis invicta* (introduced invasive
   with a painful sting)

Ribbed mussel, *Geukensia demissa* [*Modiolus demissus*]

St. Francis' satyr, *Neonympha mitchellii francisci*

Saltmarsh greenhead, *Tabanus nigrovittatus*

Sand flea (mole crab), *Emerita talpoida*

Stem borer, *Chilo demotellus*

Waccamaw fatmucket, *Lampsilis fullerkati*

Waccamaw spike, *Elliptio waccamawensis*

Water flea, *Daphnia* spp.

Fishes

American eel, *Anguilla rostrata*

American shad, *Alosa sapidissima*

Atlantic sturgeon, *Acipenser brevirostrum*

Banded killifish, *Fundulus diaphanus*

Blackbanded sunfish, *Enneacanthus chaetodon*

Bluefish, *Pomatomus saltatrix*

Bluegill, *Lepomis macrochirus*

Bowfin, *Amia calva*

Bull shark, *Caracharhinus leucas*

Cape Fear shiner, *Notropis mekistocholas*

Creek chub, *Semotilus atromaculatus*

Darter goby, *Ctenogobius boleosoma*

European eel, *Anguilla anguilla*

Great white shark, *Carcharodon carcharias*

Herring, family Clupidae

King mackerel, *Scomberomorus cavalla*

Lake trout, *Salvelinus namaycush*

Largemouth bass, *Micropterus salmoides*

Longnose gar, *Lepisosteus osseus*

Longnose sturgeon, *Acipenser oxyrhynchus*

Menhaden (mossbunker), *Brevoortia tyrannus*

Mummichog, *Fundulus heteroclitus*

Pacific salmon, any of five species of *Oncorhynchus*
    (e.g., sockeye salmon, *O. nerka*)

Pinewoods darter, *Etheostoma mariae*

Red drum [redfish], *Sciaenops ocellatus*

Redfin pickerel, *Esox americanus americanus*

Sandhills chub, *Semotilus lumbee*

Striped bass, *Morone saxatilis*

Tessellated darter, *Etheostoma olmstedi*

Waccamaw darter, *Etheostoma perlongum*

Waccamaw killifish, *Fundulus waccamensis*

Waccamaw silverside, *Menidia extensa*

Amphibians

Bullfrog, *Lithobates [Rana] catesbeianus*
Common mudpuppy, *Necturus maculosus*
Dwarf waterdog, *Necturus punctatus*
Eastern spadefoot toad, *Scaphiopus holbrookii*
Neuse River waterdog, *Necturus lewisi*
Pine barrens treefrog, *Hyla andersonii*

Reptiles

American alligator, *Alligator mississippiensis*
Burmese python, *Python molurus bivittatus*
Carolina salt marsh snake, *Nerodia sipedon williamengelsi*
Cottonmouth, *Agkistrodon piscivorus*
Diamondback terrapin, *Malaclemys terrapin*
Eastern diamondback rattlesnake, *Crotalus adamanteus*
Gopher tortoise, *Gopherus polyphemus*
Green sea turtle, *Chelonia mydas*
Hawksbill sea turtle, *Eretmochelys imbricata*
Kemp's ridley sea turtle, *Lepidochelys kempii*
Leatherback sea turtle, *Dermochelys coriacea*
Loggerhead sea turtle, *Caretta caretta*
Northern water snake, *Nerodia [Natrix] sipedon*
Pigmy rattlesnake, *Sistrurus miliarius*
Pine snake, *Pituophis melanoleucus*
Rat snake, *Elaphe obsoleta* (may occur in either a black
    or a yellow form in North Carolina)
Spotted turtle, *Clemmys guttata*
Western diamondback rattlesnake, *Crotalus atrox*

Birds

American black duck, *Anas rubripes*
American oystercatcher, *Haematopus palliatus*
American woodcock, *Scolopax [Philohela] minor*
Anhinga, *Anhinga anhinga*

Bachman's sparrow, *Peucaea [Aimophila] aestivalis*

Bachman's warbler, *Vermivora bachmanii**

Barnacle goose, *Branta leucopsis*

Black scoter, *Melanitta nigra*

Black skimmer, *Rynchops niger*

Blue-winged teal, *Anas dicors*

Bobolink (rice bird), *Dolichonyx oryzivorus*

Brown pelican, *Pelecanus occidentalis*

Canada goose, *Branta canadensis*

Carolina parakeet, *Conuropsis carolinensis**

Chimney swift, *Chaetura pelagica*

Clapper rail (marsh hen), *Rallus longirostris*

Dowitcher, *Limnodromus* spp.

Eastern bluebird, *Sialia sialis*

Eider, *Somateria* spp.

European starling, *Sturnus vulgaris* (introduced invasive)

Forster's tern, *Sterna forsteri*

Gadwall, *Anas strepera*

Goldeneye, *Bucephala* spp.

Great blue heron, *Ardea herodias*

Great egret (common egret), *Ardea alba [Casmerodius albus]*

Great horned owl, *Bubo virginianus*

Harlequin duck, *Histrionicus histrionicus*

House sparrow, *Passer domesticus*

Ivory-billed woodpecker, *Campephilus principalis**

King rail, *Rallus elegans*

King vulture, *Sarcoramphus papa*

Kirtland's warbler, *Setophaga [Dendroica] kirtlandii*

Labrador duck, *Camptorhynchus labradorius**

Least tern, *Sternula antillarum [Sterna albifrons]*

Lesser scaup (blackhead), *Aythya affinis*

Long-tailed duck (oldsquaw), *Clangula hyemalis*

Mallard, *Anas platyrhynchos*

Marsh wren (long-billed marsh wren), *Cistothorus [Telmatodytes] palustris*

Merganser, *Mergus* spp.

Mississippi kite, *Ictinia mississippiensis*

Northern harrier (marsh hawk), *Circus cyaneus*

Northern parula, *Setophaga [Parula] americana*

Northern pintail, *Anas acuta*

Osprey, *Pandion haliaetus*

Passenger pigeon, *Ectopistes migratorius**

Prairie-chicken, *Tympanuchus* spp.

Purple martin, *Progne subis*

Purple sandpiper, *Calidris [Erolia] maritima*

Red-cockaded woodpecker, *Picoides [Dendrocopos] borealis*

Red-tailed hawk, *Buteo jamaicensis*

Ring-necked pheasant, *Phasianus colchicus* (introduced)

Ruddy turnstone, *Arenaria interpres*

Seaside sparrow, *Ammodramus maritimus [Ammospiza maritima]*

Snow goose, *Anser (Chen) caerulescens* (two subspecies, greater,
A. c. atlanticus, and lesser, *A. c. caerulescens*)

Snowy owl, *Bubo [Nyctea] scandiacus*

Surf scoter, *Melanitta perspicillata*

Swainson's warbler, *Limnothlypis swainsonii*

Thick-billed parrot, *Rhynchopsitta pachyrhyncha*

Tree swallow, *Tachycineta bicolor*

Tundra swan, *Cygnus columbianus*

Wayne's black-throated green warbler, *Setophaga [Dendroica]
virens waynei*

White-winged scoter, *Melanitta fusca*

Whooping crane, *Grus americana*

Wilson's snipe, *Gallinago delicata [Capella gallinago]*

Wood stork, *Mycteria americana* (erroneously called "wood ibis")

Yellow-rumped warbler (myrtle warbler), *Setophaga
[Dendroica] coronata*

Mammals

Atlantic right whale, *Eubalaena glacialis*

Bison (buffalo), *Bos [Bison] bison*

Black bear, *Ursus americanus*

Bottlenose dolphin, *Tursiops truncatus*

Caribou, *Rangifer tarandus* (known as reindeer in Eurasia)

Coyote, *Canis latrans*

Dismal Swamp bog lemming, *Synaptomys cooperi helaletes*

Dismal Swamp short-tailed shrew, *Blarina brevicauda telmalestes*

Dismal Swamp southeastern shrew, *Sorex longirostris fisheri*

Eastern wood rat, *Neotoma floridana*

Florida manatee, *Trichechus manatus*

Giant panda, *Ailuropoda melanoleuca*

Gray whale, *Eschrichtius robustus*

Gray wolf, *Canis lupus*

Harp seal, *Pagophilus groenlandicus*

Humpback whale, *Megaptera novaeangliae*

Jaguar, *Panthera onca*

Long-finned pilot whale, *Globicephala melas*

Marsh rabbit, *Sylvilagus palustris* (notable subspecies, *S. p. hefneri*)

Marsh rice rat, *Oryzomys palustris*

Mink, *Mustela vison*

Nutria (coypu), *Myocaster coypus* (introduced invasive)

Pronghorn (antelope), *Antilocapra americana*

Raccoon, *Procyon lotor*

Rafinesque's big-eared bat, *Corynorhinus* [*Plecotus*] *rafinesquii*

Red fox, *Vulpes vulpes*

Red wolf, *Canis rufus*

River otter, *Lontra* [*Lutra*] *canadensis*

Seminole bat, *Lasiurus seminolus*

Short-finned pilot whale, *Globicephala macrorhynchus*

Southern flying squirrel, *Glaucomys volans*

Sperm whale, *Physeter macrocephalus*

Steller's sea cow, *Hydrodamalis gigas**

Swamp rabbit, *Sylvilagus aquaticus*

Waccamaw whale, *Balaenula* sp. (fossil)